More Praise for *How to Create a Mind*

"This book is a Rosetta stone for the mystery of human thought. Even more remarkably, it is a blueprint for creating artificial consciousness that is as persuasive and emotional as our own. Kurzweil deals with the subject of consciousness better than anyone from Blackmore to Dennett. His persuasive thought experiment is of Einstein quality: It forces recognition of the truth."

—Martine Rothblatt, chairman and CEO, United
Therapeutics; creator of Sirius XM Satellite Radio

"Kurzweil's book is a shining example of his prodigious ability to synthesize ideas from disparate domains and explain them to readers in simple, elegant language. Just as Chanute's *Progress in Flying Machines* ushered in the era of aviation over a century ago, this book is the harbinger of the coming revolution in artificial intelligence that will fulfill Kurzweil's own prophecies about it."

—Dileep George, AI scientist; pioneer of hierarchical
models of the neocortex; cofounder of Numenta
and Vicarious Systems

"Ray Kurzweil's understanding of the brain and artificial intelligence will dramatically impact every aspect of our lives, every industry on Earth, and how we think about our future. If you care about any of these, *read* this book!"

—Peter H. Diamandis, chairman and CEO,
X PRIZE; executive chairman, Singularity
University; author of the *New York Times* bestseller
Abundance: The Future Is Better Than You Think

HOW TO CREATE A MIND

HOW TO CREATE A MIND

THE SECRET OF HUMAN THOUGHT REVEALED

RAY KURZWEIL

VIKING

VIKING
Published by the Penguin Group
Penguin Group (USA) Inc., 375 Hudson Street, New York, New York 10014, U.S.A. • Penguin Group (Canada), 90 Eglinton Avenue East, Suite 700, Toronto, Ontario, Canada M4P 2Y3 (a division of Pearson Penguin Canada Inc.) • Penguin Books Ltd, 80 Strand, London WC2R 0RL, England • Penguin Ireland, 25 St. Stephen's Green, Dublin 2, Ireland (a division of Penguin Books Ltd) • Penguin Group (Australia), 707 Collins Street, Melbourne, Victoria 3008, Australia (a division of Pearson Australia Group Pty Ltd) • Penguin Books India Pvt Ltd, 11 Community Centre, Panchsheel Park, New Delhi–110 017, India • Penguin Group (NZ), 67 Apollo Drive, Rosedale, Auckland 0632, New Zealand (a division of Pearson New Zealand Ltd) • Penguin Books, Rosebank Office Park, 181 Jan Smuts Avenue, Parktown North 2193, South Africa • Penguin China, B7 Jaiming Center, 27 East Third Ring Road North, Chaoyang District, Beijing 100020, China

Penguin Books Ltd, Registered Offices:
80 Strand, London WC2R 0RL, England

First published in 2012 by Viking Penguin,
a member of Penguin Group (USA) Inc.

1   3   5   7   9   10   8   6   4   2

"Red" by Amoo Oluseun. Used by permission of the author.
"The picture's pretty bleak, gentlemen . . ." from *The Far Side* by Gary Larson (November 7, 1985). Used by permission of Creators Syndicate.

Illustration credits
Page 10: Created by Wolfgang Beyer (Creative Commons Attribution–Share Alike 3.0 License). 21: Photo by Timeline (Creative Commons Attribution–Share Alike 3.0 License). 84 (two figures): From "The Geometric Structure of the Brain Fiber Pathways," by Van J. Wedeen, Douglas L. Rosene, Ruopene Wang, Guangping Dai, Farzad Mortazavi, Patric Hagmann, Jon H. Kaas, and Wen-Yih I. Tseng, *Science,* March 30, 2012. Reprinted with permission of AAAS (American Association for the Advancement of Science). 85: Photo provided by Yeatesh (Creative Commons Attribution–Share Alike 3.0 License). 134 (two): Images by Marvin Minsky. Used by permission of Marvin Minsky. Some credits appear adjacent to the respective images. Other images designed by Ray Kurzweil, illustrated by Laksman Frank.

Library of Congress Cataloging-in-Publication Data

Kurzweil, Ray.
How to create a mind : the secret of human thought revealed / Ray Kurzweil.
p.   cm.
Includes bibliographical references and index.
ISBN 978-0-670-02529-9
1. Brain—Localization of functions. 2. Self-consciousness (Awareness)
3. Artificial intelligence. I. Title.
QP385.K87 2012
612.8'2—dc23   2012027185

Printed in the United States of America
Set in Minion Pro with DIN
Designed by Daniel Lagin

To Leo Oscar Kurzweil. You are entering an extraordinary world.

ACKNOWLEDGMENTS

I'd like to express my gratitude to my wife, Sonya, for her loving patience through the vicissitudes of the creative process;

To my children, Ethan and Amy; my daughter-in-law, Rebecca; my sister, Enid; and my new grandson, Leo, for their love and inspiration;

To my mother, Hannah, for supporting my early ideas and inventions, which gave me the freedom to experiment at a young age, and for keeping my father alive during his long illness;

To my longtime editor at Viking, Rick Kot, for his leadership, steady and insightful guidance, and expert editing;

To Loretta Barrett, my literary agent for twenty years, for her astute and enthusiastic guidance;

To Aaron Kleiner, my long-term business partner, for his devoted collaboration for the past forty years;

To Amara Angelica for her devoted and exceptional research support;

To Sarah Black for her outstanding research insights and ideas;

To Laksman Frank for his excellent illustrations;

To Sarah Reed for her enthusiastic organizational support;

To Nanda Barker-Hook for her expert organization of my public events on this and other topics;

ACKNOWLEDGMENTS

To Amy Kurzweil for her guidance on the craft of writing;

To Cindy Mason for her research support and ideas on AI and the mind-body connection;

To Dileep George for his discerning ideas and insightful discussions by e-mail and otherwise;

To Martine Rothblatt for her dedication to all of the technologies I discuss in the book and for our collaborations in developing technologies in these areas;

To the KurzweilAI.net team, who provided significant research and logistical support for this project, including Aaron Kleiner, Amara Angelica, Bob Beal, Casey Beal, Celia Black-Brooks, Cindy Mason, Denise Scutellaro, Joan Walsh, Giulio Prisco, Ken Linde, Laksman Frank, Maria Ellis, Nanda Barker-Hook, Sandi Dube, Sarah Black, Sarah Brangan, and Sarah Reed;

To the dedicated team at Viking Penguin for all of their thoughtful expertise, including Clare Ferraro (president), Carolyn Coleburn (director of publicity), Yen Cheong and Langan Kingsley (publicists), Nancy Sheppard (director of marketing), Bruce Giffords (production editor), Kyle Davis (editorial assistant), Fabiana Van Arsdell (production director), Roland Ottewell (copy editor), Daniel Lagin (designer), and Julia Thomas (jacket designer);

To my colleagues at Singularity University for their ideas, enthusiasm, and entrepreneurial energy;

To my colleagues who have provided inspired ideas reflected in this volume, including Barry Ptolemy, Ben Goertzel, David Dalrymple, Dileep George, Felicia Ptolemy, Francis Ganong, George Gilder, Larry Janowitch, Laura Deming, Lloyd Watts, Martine Rothblatt, Marvin Minsky, Mickey Singer, Peter Diamandis, Raj Reddy, Terry Grossman, Tomaso Poggio, and Vlad Sejnoha;

To my peer expert readers, including Ben Goertzel, David Gamez, Dean Kamen, Dileep George, Douglas Katz, Harry George, Lloyd Watts,

ACKNOWLEDGMENTS

Martine Rothblatt, Marvin Minsky, Paul Linsay, Rafael Reif, Raj Reddy, Randal Koene, Dr. Stephen Wolfram, and Tomaso Poggio;

To my in-house and lay readers whose names appear above;

And, finally, to all of the creative thinkers in the world who inspire me every day.

CONTENTS

HOW TO CREATE A MIND

INTRODUCTION

The Brain—is wider than the Sky—
For—put them side by side—
The one the other will contain
With ease—and You—beside—
The Brain is deeper than the sea—
For—hold them—Blue to Blue—
The one the other will absorb—
As Sponges—Buckets—do—
The Brain is just the weight of God—
For—Heft them—Pound for Pound—
And they will differ—if they do—
As Syllable from Sound

—Emily Dickinson

As the most important phenomenon in the universe, intelligence is capable of transcending natural limitations, and of transforming the world in its own image. In human hands, our intelligence has enabled us to overcome the restrictions of our biological heritage and to change ourselves in the process. We are the only species that does this.

The story of human intelligence starts with a universe that is capable of encoding information. This was the enabling factor that allowed evolution to take place. How the universe got to be this way is itself an interesting story. The standard model of physics has dozens of constants that need to be precisely what they are, or atoms would not have been possible, and there would have been no stars, no planets, no brains, and no books on brains. That the laws of physics are so precisely tuned to have allowed the evolution of information appears to be incredibly unlikely. Yet by the anthropic principle, we would not be talking about it if it were not the case. Where some people see a divine hand, others see a multiverse spawning an evolution of universes with the boring (non-information-bearing) ones dying out. But regardless of how our universe got to be the way it is, we can start our story with a world based on information.

The story of evolution unfolds with increasing levels of abstraction. Atoms—especially carbon atoms, which can create rich information structures by linking in four different directions—formed increasingly complex molecules. As a result, physics gave rise to chemistry.

A billion years later, a complex molecule called DNA evolved, which could precisely encode lengthy strings of information and generate organisms described by these "programs." As a result, chemistry gave rise to biology.

At an increasingly rapid rate, organisms evolved communication and decision networks called nervous systems, which could coordinate the increasingly complex parts of their bodies as well as the behaviors that facilitated their survival. The neurons making up nervous systems aggregated into brains capable of increasingly intelligent behaviors. In this way, biology gave rise to neurology, as brains were now the cutting edge of storing and manipulating information. Thus we went from atoms to molecules to DNA to brains. The next step was uniquely human.

The mammalian brain has a distinct aptitude not found in any other class of animal. We are capable of *hierarchical* thinking, of understanding a structure composed of diverse elements arranged in a pattern, representing that arrangement with a symbol, and then using that symbol as an element in a yet more elaborate configuration. This capability takes place in a brain structure called the neocortex, which in humans has achieved a threshold of sophistication and capacity such that we are able to call these patterns *ideas*. Through an unending recursive process we are capable of building ideas that are ever more complex. We call this vast array of recursively linked ideas *knowledge*. Only *Homo sapiens* have a knowledge base that itself evolves, grows exponentially, and is passed down from one generation to another.

Our brains gave rise to yet another level of abstraction, in that we have used the intelligence of our brains plus one other enabling factor, an opposable appendage—the thumb—to manipulate the environment to build tools. These tools represented a new form of evolution, as neurology gave rise to technology. It is only because of our tools that our knowledge base has been able to grow without limit.

Our first invention was the story: spoken language that enabled us to represent ideas with distinct utterances. With the subsequent invention of written language we developed distinct shapes to symbolize our ideas. Libraries of written language vastly extended the ability of our unaided brains to retain and extend our knowledge base of recursively structured ideas.

There is some debate as to whether other species, such as chimpanzees, have the ability to express hierarchical ideas in language. Chimps are capable of learning a limited set of sign language symbols, which they can use to communicate with human trainers. It is clear, however, that there are distinct limits to the complexity of the knowledge structures with which chimps are capable of dealing. The sentences that they can express are limited to specific simple noun-verb sequences and are not capable of the indefinite expansion of complexity characteristic of humans. For an entertaining example of the complexity of human-generated language, just read one of the spectacular multipage-length sentences in a Gabriel García Márquez story or novel—his six-page story "The Last Voyage of the Ghost" is a single sentence and works quite well in both Spanish and the English translation.[1]

The primary idea in my three previous books on technology (*The Age of Intelligent Machines,* written in the 1980s and published in 1989; *The Age of Spiritual Machines,* written in the mid- to late 1990s and published in 1999; and *The Singularity Is Near,* written in the early 2000s and published in 2005) is that an evolutionary process inherently accelerates (as a result of its increasing levels of abstraction) and that its products grow exponentially in complexity and capability. I call this phenomenon the law of accelerating returns (LOAR), and it pertains to both biological and technological evolution. The most dramatic example of the LOAR is the remarkably predictable exponential growth in the capacity and price/performance of information technologies. The evolutionary process of technology led invariably to the computer, which has in turn enabled a vast expansion of our knowledge base, permitting extensive links from one area of knowledge to another. The Web is itself a powerful and apt example of the ability of a hierarchical system to encompass a vast array of knowledge while preserving its inherent structure. The world itself is inherently hierarchical— trees contain branches; branches contain leaves; leaves contain veins. Buildings contain floors; floors contain rooms; rooms contain doorways, windows, walls, and floors.

We have also developed tools that are now enabling us to understand our own biology in precise information terms. We are rapidly reverse-engineering the information processes that underlie biology, including that of our brains. We now possess the object code of life in the form of the human genome, an achievement that was itself an outstanding example of exponential growth, in that the amount of genetic data the world has sequenced has approximately doubled every year for the past twenty years.[2] We now have the ability to simulate on computers how sequences of base pairs give rise to sequences of amino acids that fold up into three-dimensional proteins, from which all of biology is constructed. The complexity of proteins for which we can simulate protein folding has been steadily increasing as computational resources continue to grow exponentially.[3] We can also simulate how proteins interact with one another in an intricate three-dimensional dance of atomic forces. Our growing understanding of biology is one important facet of discovering the intelligent secrets that evolution has bestowed on us and then using these biologically inspired paradigms to create ever more intelligent technology.

There is now a grand project under way involving many thousands of scientists and engineers working to understand the best example we have of an intelligent process: the human brain. It is arguably the most important effort in the history of the human-machine civilization. In *The Singularity Is Near* I made the case that one corollary of the law of accelerating returns is that other intelligent species are likely not to exist. To summarize the argument, if they existed we would have noticed them, given the relatively brief time that elapses between a civilization's possessing crude technology (consider that in 1850 the fastest way to send nationwide information was the Pony Express) to its possessing technology that can transcend its own planet.[4] From this perspective, reverse-engineering the human brain may be regarded as the most important project in the universe.

The goal of the project is to understand precisely how the human brain works, and then to use these revealed methods to better understand ourselves, to fix the brain when needed, and—most relevant to the subject of this book—to create even more intelligent machines. Keep in mind that greatly amplifying a natural phenomenon is precisely what engineering is capable of doing. As an example, consider the rather subtle phenomenon of Bernoulli's principle, which states that there is slightly less air pressure over a moving curved surface than over a moving flat one. The mathematics of how Bernoulli's principle produces wing lift is still not yet fully settled among scientists, yet engineering has taken this delicate insight, focused its powers, and created the entire world of aviation.

In this book I present a thesis I call the pattern recognition theory of mind (PRTM), which, I argue, describes the basic algorithm of the neocortex (the region of the brain responsible for perception, memory, and critical thinking). In the chapters ahead I describe how recent neuro-science research, as well as our own thought experiments, leads to the inescapable conclusion that this method is used consistently across the neocortex. The implication of the PRTM combined with the LOAR is that we will be able to engineer these principles to vastly extend the powers of our own intelligence.

Indeed this process is already well under way. There are hundreds of tasks and activities formerly the sole province of human intelligence that can now be conducted by computers, usually with greater precision and at a vastly greater scale. Every time you send an e-mail or connect a cell phone call, intelligent algorithms optimally route the information. Obtain an electrocardiogram, and it comes back with a computer diagnosis that rivals that of doctors. The same is true for blood cell images. Intelligent algorithms automatically detect credit card fraud, fly and land airplanes, guide intelligent weapons systems, help design products with intelligent computer-aided design, keep track of just-in-time inventory levels, assemble products in robotic factories, and play games such as chess and even the subtle game of Go at master levels.

Millions of people witnessed the IBM computer named Watson play the natural-language game of *Jeopardy!* and obtain a higher score than the best two human players in the world combined. It should be noted that not only did Watson read and "understand" the subtle language in the *Jeopardy!* query (which includes such phenomena as puns and metaphors), but it obtained the knowledge it needed to come up with a response from understanding hundreds of millions of pages of natural-language documents including Wikipedia and other encyclopedias on its own. It needed to master virtually every area of human intellectual endeavor, including history, science, literature, the arts, culture, and more. IBM is now working with Nuance Speech Technologies (formerly Kurzweil Computer Products, my first company) on a new version of Watson that will read medical literature (essentially all medical journals and leading medical blogs) to become a master diagnostician and medical consultant, using Nuance's clinical language–understanding technologies. Some observers have argued that Watson does not really "understand" the *Jeopardy!* queries or the encyclopedias it has read because it is just engaging in "statistical analysis." A key point I will describe here is that the mathematical techniques that have evolved in the field of artificial intelligence (such as those used in Watson and Siri, the iPhone assistant) are mathematically very similar to the methods that biology evolved in the form of the neocortex. If understanding language and other phenomena through statistical analysis does not count as true understanding, then humans have no understanding either.

Watson's ability to intelligently master the knowledge in natural-language documents is coming to a search engine near you, and soon. People are already talking to their phones in natural language (via Siri, for example, which was also contributed to by Nuance). These natural-language assistants will rapidly become more intelligent as they utilize more of the Watson-like methods and as Watson itself continues to improve.

The Google self-driving cars have logged 200,000 miles in the busy cities and towns of California (a figure that will undoubtedly be much higher by the time this book hits the real and virtual shelves). There are many other examples of artificial intelligence in today's world, and a great deal more on the horizon.

As further examples of the LOAR, the spatial resolution of brain scanning and the amount of data we are gathering on the brain are doubling every year. We are also demonstrating that we can turn this data into working models and simulations of brain regions. We have succeeded in reverse-engineering key functions of the auditory cortex, where we process information about sound; the visual cortex, where we process information from our sight; and the cerebellum, where we do a portion of our skill formation (such as catching a fly ball).

The cutting edge of the project to understand, model, and simulate the human brain is to reverse-engineer the cerebral neocortex, where we do our recursive hierarchical thinking. The cerebral cortex, which accounts for 80 percent of the human brain, is composed of a highly repetitive structure, allowing humans to create arbitrarily complex structures of ideas.

In the pattern recognition theory of mind, I describe a model of how the human brain achieves this critical capability using a very clever structure designed by biological evolution. There are details in this cortical mechanism that we do not yet fully understand, but we know enough about the functions it needs to perform that we can nonetheless design algorithms that accomplish the same purpose. By beginning to understand the neocortex, we are now in a position to greatly amplify its powers, just as the world of aviation has vastly amplified the powers of Bernoulli's principle. The operating principle of the neocortex is arguably the most important idea in the world, as it is capable of representing all knowledge and skills as well as creating new knowledge. It is the neocortex, after all, that has been responsible for every novel, every song, every painting, every scientific discovery, and the multifarious other products of human thought.

There is a great need in the field of neuroscience for a theory that ties together the extremely disparate and extensive observations that are being reported on a daily basis. A unified theory is a crucial requirement in every major area of science. In chapter 1 I'll describe how two daydreamers unified biology and physics, fields that had previously seemed hopelessly disordered and varied, and then address how such a theory can be applied to the landscape of the brain.

Today we often encounter great celebrations of the complexity of the human brain. Google returns some 30 million links for a search request for quotations on that topic. (It is impossible to translate this into the number of actual quotations it is returning, however, as some of the Web sites linked have multiple quotes, and some have none.) James D. Watson himself wrote in 1992 that "the brain is the last and grandest biological frontier, the most complex thing we have yet discovered in our universe." He goes on to explain why he believes that "it contains hundreds of billions of cells interlinked through trillions of connections. The brain boggles the mind."[5]

I agree with Watson's sentiment about the brain's being the grandest biological frontier, but the fact that it contains many billions of cells and trillions of connections does not necessarily make its primary method complex if we can identify readily understandable (and re-creatable) patterns in those cells and connections, especially massively redundant ones.

Let's think about what it means to be complex. We might ask, is a forest complex? The answer depends on the perspective you choose to take. You could note that there are many thousands of trees in the forest and that each one is different. You could then go on to note that each tree has thousands of branches and that each branch is completely different. Then you could proceed to describe the convoluted vagaries of a single branch. Your conclusion might be that the forest has a complexity beyond our wildest imagination.

But such an approach would literally be a failure to see the forest for the trees. Certainly there is a great deal of fractal variation among trees and branches, but to correctly understand the principles of a forest you would do better to start by identifying the distinct patterns of redundancy with stochastic (that is, random) variation that are found there. It would be fair to say that the concept of a forest is simpler than the concept of a tree.

Thus it is with the brain, which has a similar enormous redundancy, especially in the neocortex. As I will describe in this book, it would be fair to say that there is more complexity in a single neuron than in the overall structure of the neocortex.

My goal in this book is definitely not to add another quotation to the millions that already exist attesting to how complex the brain is, but rather to impress you with the power of its simplicity. I will do so by describing how a basic ingenious mechanism for recognizing, remembering, and predicting a pattern, repeated in the neocortex hundreds of millions of times, accounts for the great diversity of our thinking. Just as an astonishing diversity of organisms arises from the different combinations of the values of the genetic code found in nuclear and mitochondrial DNA, so too does an astounding array of ideas, thoughts, and skills form based on the values of the patterns (of connections and synaptic strengths) found in and between our neocortical pattern recognizers. As MIT neuroscientist Sebastian Seung says, "Identity lies not in our genes, but in the connections between our brain cells."[6]

We need to distinguish between true complexity of design and apparent complexity. Consider the famous Mandelbrot set, the image of which has long been a symbol of complexity. To appreciate its apparent complication, it is useful to zoom in on its image (which you can access via the links in this endnote).[7] There is endless intricacy within intricacy, and they are always different. Yet the design—the formula—for the Mandelbrot set couldn't be simpler. It is six characters long: $Z = Z^2 + C$, in which Z is a "complex" number (meaning a pair of numbers) and C is a constant. It is

One view of the display of the Mandelbrot set, a simple formula that is iteratively applied. As one zooms in on the display, the images constantly change in apparently complex ways.

not necessary to fully understand the Mandelbrot function to see that it is simple. This formula is applied iteratively and at every level of a hierarchy. The same is true of the brain. Its repeating structure is not as simple as that of the six-character formula of the Mandelbrot set, but it is not nearly as complex as the millions of quotations on the brain's complexity would suggest. This neocortical design is repeated over and over at every level of the conceptual hierarchy represented by the neocortex. Einstein articulated my goals in this book well when he said that "any intelligent fool can make things bigger and more complex . . . but it takes . . . a lot of courage to move in the opposite direction."

So far I have been talking about the brain. But what about the mind? For example, how does a problem-solving neocortex attain consciousness? And while we're on the subject, just how many conscious minds do we have in our brain? There is evidence that suggests there may be more than one.

Another pertinent question about the mind is, what is free will, and do we have it? There are experiments that appear to show that we start implementing our decisions before we are even aware that we have made them. Does that imply that free will is an illusion?

Finally, what attributes of our brain are responsible for forming our identity? Am I the same person I was six months ago? Clearly I am not exactly the same as I was then, but do I have the same identity?

We'll review what the pattern recognition theory of mind implies about these age-old questions.

CHAPTER 1

THOUGHT EXPERIMENTS ON THE WORLD

Darwin's theory of natural selection came very late in the history of thought.

Was it delayed because it opposed revealed truth, because it was an entirely new subject in the history of science, because it was characteristic only of living things, or because it dealt with purpose and final causes without postulating an act of creation? I think not. Darwin simply discovered the role of selection, a kind of causality very different from the push-pull mechanisms of science up to that time. The origin of a fantastic variety of living things could be explained by the contribution of which novel features, possibly of random provenance, made it to survival. There was little or nothing in physical or biological science that foreshadowed selection as a causal principle.

—B. F. Skinner

Nothing is at last sacred but the integrity of your own mind.

—Ralph Waldo Emerson

A Metaphor from Geology

In the early nineteenth century geologists pondered a fundamental question. Great caverns and canyons such as the Grand Canyon in the United States and Vikos Gorge in Greece (reportedly the deepest canyon in the world) existed all across the globe. How did these majestic formations get there?

Invariably there was a stream of water that appeared to take advantage of the opportunity to course through these natural structures, but prior to the mid-nineteenth century, it had seemed absurd that these gentle flows could be the creator of such huge valleys and cliffs. British geologist Charles Lyell (1797–1875), however, proposed that it was indeed the movement of water that had carved out these major geological modifications over great periods of time, essentially one grain of rock at a time. This proposal was initially met with ridicule, but within two decades Lyell's thesis achieved mainstream acceptance.

One person who was carefully watching the response of the scientific community to Lyell's radical thesis was English naturalist Charles Darwin (1809–1882). Consider the situation in biology around 1850. The field was endlessly complex, faced with countless species of animals and plants, any one of which presented great intricacy. If anything, most scientists resisted any attempt to provide a unifying theory of nature's dazzling variation. This diversity served as a testament to the glory of God's creation, not to mention to the intelligence of the scientists who were capable of mastering it.

Darwin approached the problem of devising a general theory of species by making an analogy with Lyell's thesis to account for the gradual changes in the features of species over many generations. He combined this insight with his own thought experiments and observations in his famous *Voyage of the Beagle*. Darwin argued that in each generation the individuals that could best survive in their ecological niche would be the individuals to create the next generation.

On November 22, 1859, Darwin's book *On the Origin of Species* went on sale, and in it he made clear his debt to Lyell:

> I am well aware that this doctrine of natural selection, exemplified in the above imaginary instances, is open to the same objections which were at first urged against Sir Charles Lyell's noble views on "the modern changes of the earth, as illustrative of geology"; but we now very seldom hear the action, for instance, of the coast-waves called a trifling and insignificant cause, when applied to the excavation of gigantic valleys or to the formation of the longest lines of inland cliffs. Natural selection can act only by the preservation and accumulation of infinitesimally small inherited modifications, each profitable to the preserved being; and as modern geology has almost banished such views as the excavation of a great valley by a single diluvial wave, so will natural selection, if it be a true principle, banish the belief of the continued creation of new organic beings, or of any great and sudden modification in their structure.[1]

Charles Darwin, author of *On the Origin of Species,* which established the idea of biological evolution.

There are always multiple reasons why big new ideas are resisted, and it is not hard to identify them in Darwin's case. That we were descended not from God but from monkeys, and before that, worms, did not sit well with many commentators. The implication that our pet dog was our cousin, as was the caterpillar, not to mention the plant it walked on (a millionth or billionth cousin, perhaps, but still related), seemed a blasphemy to many.

But the idea quickly caught on because it brought coherence to what had previously been a plethora of apparently unrelated observations. By 1872, with the publication of the sixth edition of *On the Origin of Species*, Darwin added this passage: "As a record of a former state of things, I have retained in the foregoing paragraphs . . . several sentences which imply that naturalists believe in the separate creation of each species; and I have been much censured for having thus expressed myself. But undoubtedly this was the general belief when the first edition of the present work appeared. . . . Now things are wholly changed, and almost every naturalist admits the great principle of evolution."[2]

Over the next century Darwin's unifying idea deepened. In 1869, only a decade after the original publication of *On the Origin of Species*, Swiss physician Friedrich Miescher (1844–1895) discovered a substance he called "nuclein" in the cell nucleus, which turned out to be DNA.[3] In 1927 Russian biologist Nikolai Koltsov (1872–1940) described what he called a "giant hereditary molecule," which he said was composed of "two mirror strands that would replicate in a semi-conservative fashion using each strand as a template." His finding was also condemned by many. The communists considered it to be fascist propaganda, and his sudden, unexpected death has been attributed to the secret police of the Soviet Union.[4] In 1953, nearly a century after the publication of Darwin's seminal book, American biologist James D. Watson (born in 1928) and English biologist Francis Crick (1916–2004) provided the first accurate characterization of the structure of DNA, describing it as a double helix of two long twisting molecules.[5] It is worth pointing out that their finding was based on what is now known as "photo 51," taken by their colleague Rosalind Franklin using

Rosalind Franklin took the critical picture of DNA (using X-ray crystallography) that enabled Watson and Crick to accurately describe the structure of DNA for the first time.

X-ray crystallography, which was the first representation that showed the double helix. Given the insights derived from Franklin's image, there have been suggestions that she should have shared in Watson and Crick's Nobel Prize.[6]

With the description of a molecule that could code the program of biology, a unifying theory of biology was now firmly in place. It provided a simple and elegant foundation to all of life. Depending only on the values of the base pairs that make up the DNA strands in the nucleus (and to a lesser degree the mitochondria), an organism would mature into a blade of grass or a human being. This insight did not eliminate the delightful diversity of nature, but we now understand that the extraordinary diversity of nature stems from the great assortment of structures that can be coded on this universal molecule.

Riding on a Light Beam

At the beginning of the twentieth century the world of physics was upended through another series of thought experiments. In 1879 a boy was born to a German engineer and a housewife. He didn't start to talk until the age of three and was reported to have had problems in school at the age of nine. At sixteen he was daydreaming about riding on a moonbeam.

This young boy was aware of English mathematician Thomas Young's (1773–1829) experiment in 1803 that established that light is composed of waves. The conclusion at that time was that light waves must be traveling through some sort of medium; after all, ocean waves traveled through water and sound waves traveled through air and other materials. Scientists called the medium through which light waves travel the "ether." The boy was also aware of the 1887 experiment by American scientists Albert Michelson (1852–1931) and Edward Morley (1838–1923) that attempted to confirm the existence of the ether. That experiment was based on the analogy of traveling in a rowboat up- and downstream in a river. If you are paddling at a fixed speed, then your speed as measured from the shore will be faster if you are paddling with the stream as opposed to going against it. Michelson and Morley assumed that light would travel through the ether at a constant speed (that is, at the speed of light). They reasoned that the speed of sunlight when Earth is traveling toward the sun in its orbit (as measured from our vantage point on Earth) versus its apparent speed when Earth is traveling away from the sun must be different (by twice the speed of Earth). Proving that would confirm the existence of the ether. However, what they discovered was that there was no difference in the speed of the sunlight passing Earth regardless of where Earth was in its orbit. Their findings disproved the idea of the "ether," but what was really going on? This remained a mystery for almost two decades.

As this German teenager imagined riding alongside a light wave, he reasoned that he should be seeing the light waves frozen, in the same way that a train would appear not to be moving if you rode alongside it at the same speed as the train. Yet he realized that this was impossible, because the speed of light is supposed to be constant regardless of your own movement. So he imagined instead riding alongside the light beam but at a somewhat slower speed. What if he traveled at 90 percent of the speed of light? If light beams are like trains, he reasoned, then he should see the light beam traveling ahead of him at 10 percent of the speed of light. Indeed, that would have to be what observers on Earth would see. But we know that the speed of light is a constant, as the Michelson-Morley experiment had shown. Thus he would necessarily see the light beam traveling ahead of him at the full speed of light. This seemed like a contradiction—how could it be possible?

The answer became evident to the German boy, whose name, incidentally, was Albert Einstein (1879–1955), by the time he turned twenty-six. Obviously—to young Master Einstein—*time itself must have slowed down for him*. He explains his reasoning in a paper published in 1905.[7] If observers on Earth were to look at the young man's watch they would see it ticking ten times slower. Indeed, when he got back to Earth, his watch would show that only 10 percent as much time had passed (ignoring, for the moment, acceleration and deceleration). From his perspective, however, his watch was ticking normally and the light beam next to him was traveling at the speed of light. The ten-times slowdown in the speed of time itself (relative to clocks on Earth) fully explains the apparent discrepancies in perspective. In the extreme, the slowdown in the passage of time would reach zero once the speed of travel reached the speed of light; hence it was impossible to ride along with the light beam. Although it was impossible to travel at the speed of light, it turned out not to be theoretically impossible to move *faster* than the light beam. Time would then move backward.

This resolution seemed absurd to many early critics. How could time itself slow down, based only on someone's speed of movement? Indeed, for eighteen years (from the time of the Michelson-Morley experiment), other thinkers had been unable to see a conclusion that was so obvious to Master Einstein. The many others who had considered this problem through the latter part of the nineteenth century had essentially "fallen off the horse" in terms of following through on the implications of a principle, sticking instead to their preconceived notions of how reality must work. (I should probably change that metaphor to "fallen off the light beam.")

Einstein's second mind experiment was to consider himself and his brother flying through space. They are 186,000 miles apart. Einstein wants to move faster but he also desires to keep the distance between them the same. So he signals his brother with a flashlight each time he wants to accelerate. Since he knows that it will take one second for the signal to reach his brother, he waits a second (after sending the signal) to initiate his own acceleration. Each time the brother receives the signal he immediately accelerates. In this way the two brothers accelerate at exactly the same time and therefore remain a constant distance apart.

But now consider what we would see if we were standing on Earth. If the brothers were moving away from us (with Albert in the lead), it would appear to take less than a second for the light to reach the brother, because he is traveling toward the light. Also we would see Albert's brother's clock as slowing down (as his speed increases as he is closer to us). For both of these reasons we would see the two brothers getting closer and closer and eventually colliding. Yet from the perspective of the two brothers, they remain a constant 186,000 miles apart.

How can this be? The answer—*obviously*—is that distances contract parallel to the motion (but not perpendicular to it). So the two Einstein brothers are getting shorter (assuming they are flying headfirst) as they get faster. This bizarre conclusion probably lost Einstein more early fans than the difference in the passage of time.

During the same year, Einstein considered the relationship of matter and energy with yet another mind experiment. Scottish physicist James Clerk Maxwell had shown in the 1850s that particles of light called photons had no mass but nonetheless carried momentum. As a child I had a device called a Crookes radiometer,[8] which consisted of an airtight glass bulb that contained a partial vacuum and a set of four vanes that rotated on a spindle. The vanes were white on one side and black on the other. The white side of each vane reflected light, and the black side absorbed light. (That's why it is cooler to wear a white T-shirt on a hot day than a black one.) When a light was shined on the device, the vanes rotated, with the dark sides moving away from the light. This is a direct demonstration that photons carry enough momentum to actually cause the vanes of the radiometer to move.[9]

The issue that Einstein struggled with is that momentum is a function of mass: Momentum is equal to mass times velocity. Thus a locomotive traveling at 30 miles per hour has a lot more momentum than, say, an insect traveling at the same speed. How, then, could there be positive momentum for a particle with zero mass?

Einstein's mind experiment consisted of a box floating in space. A

A Crookes radiometer—the vane with four wings rotates when light shines on it.

photon is emitted inside the box from the left toward the right side. The total momentum of the system needs to be conserved, so the box would have to recoil to the left when the photon was emitted. After a certain amount of time, the photon collides with the right side of the box, transferring its momentum back to the box. The total momentum of the system is again conserved, so the box now stops moving.

So far so good. But consider the perspective from the vantage point of Mr. Einstein, who is watching the box from the outside. He does not see any outside influence on the box: No particles—with or without mass—hit it, and nothing leaves it. Yet Mr. Einstein, according to the scenario above, sees the box move temporarily to the left and then stop. According to our analysis, each photon should permanently move the box to the left. Since there have been no external effects on the box or from the box, its center of mass must remain in the same place. Yet the photon inside the box, which moves from left to right, cannot change the center of mass, because it has no mass.

Or does it? Einstein's conclusion was that since the photon clearly has energy, and has momentum, it must also have a mass equivalent. The energy of the moving photon is entirely equivalent to a moving mass. We can compute what that equivalence is by recognizing that the center of mass of the system must remain stationary during the movement of the photon. Working out the math, Einstein showed that mass and energy are equivalent and are related by a simple constant. However, there was a catch: The constant might be simple, but it turned out to be enormous; it was the speed of light squared (about 1.7×10^{17} meters^2 per second^2—that is, 17 followed by 16 zeroes). Hence we get Einstein's famous $E = mc^2$.[10] Thus one ounce (28 grams) of mass is equivalent to 600,000 tons of TNT. Einstein's letter of August 2, 1939, to President Roosevelt informing him of the potential for an atomic bomb based on this formula ushered in the atomic age.[11]

You might think that this should have been obvious earlier, given that experimenters had noticed that the mass of radioactive substances decreased as a result of radiation over time. It was assumed, however, that radioactive substances contained a special high-energy fuel of some sort that was burning off. That assumption is not all wrong; it's just that the fuel that was being "burned off" was simply mass.

There are several reasons why I have opened this book with Darwin's and Einstein's mind experiments. First of all, they show the extraordinary power of the human brain. Without any equipment at all other than a pen and paper to draw the stick figures in these simple mind experiments and to write down the fairly simple equations that result from them, Einstein was able to overthrow the understanding of the physical world that dated back two centuries, deeply influence the course of history (including World War II), and usher in the nuclear age.

It is true that Einstein relied on a few experimental findings of the nineteenth century, although these experiments also did not use sophisticated equipment. It is also true that subsequent experimental validation of Einstein's theories has used advanced technologies, and if these had not been developed we would not have the validation that we possess today that Einstein's ideas are authentic and significant. However, such factors do not detract from the fact that these famous thought experiments reveal the power of human thinking at its finest.

Einstein is widely regarded as the leading scientist of the twentieth century (and Darwin would be a good contender for that honor in the nineteenth century), yet the mathematics underlying his theories is ultimately not very complicated. The thought experiments themselves were straightforward. We might wonder, then, in what respect could Einstein be considered particularly smart. We'll discuss later exactly what it was that he was doing with his brain when he came up with his theories, and where that quality resides.

Conversely, this history also demonstrates the limitations of human thinking. Einstein was able to ride his light beam without falling off (albeit he concluded that it was impossible to actually ride a light beam), but how many thousands of other observers and thinkers were completely unable to think through these remarkably uncomplicated exercises? One common failure is the difficulty that most people have in discarding and transcending the ideas and perspectives of their peers. There are other inadequacies as well, which we will discuss in more detail after we have examined how the neocortex works.

A Unified Model of the Neocortex

The most important reason I am sharing what are perhaps the most famous thought experiments in history is as an introduction to using the same approach with respect to the brain. As you will see, we can get remarkably far in figuring out how human intelligence works through some simple mind experiments of our own. Considering the subject matter involved, mind experiments should be a very appropriate approach.

If a young man's idle thoughts and the use of no equipment other than pen and paper were sufficient to revolutionize our understanding of physics, then we should be able to make reasonable progress with a phenomenon with which we are much more familiar. After all, we experience our thinking every moment of our waking lives—and our dreaming lives as well.

After we construct a model of how thinking works through this process of self-reflection, we'll examine to what extent we can confirm it through the latest observations of actual brains and the state of the art in re-creating these processes in machines.

CHAPTER 2

THOUGHT EXPERIMENTS ON THINKING

I very rarely think in words at all. A thought comes, and I may try to express it in words afterwards.

—Albert Einstein

The brain is a three-pound mass you can hold in your hand that can conceive of a universe a hundred billion light years across.

—Marian Diamond

What seems astonishing is that a mere three-pound object, made of the same atoms that constitute everything else under the sun, is capable of directing virtually everything that humans have done: flying to the moon and hitting seventy home runs, writing *Hamlet* and building the Taj Mahal—even unlocking the secrets of the brain itself.

—Joel Havemann

I started thinking about thinking around 1960, the same year that I discovered the computer. You would be hard pressed today to find a twelve-year-old who does not use a computer, but back then there were only a handful of them in my hometown of New York City. Of course these early devices did not fit in your hand, and the first one I got access to took up a large room. In the early 1960s I did some programming on an IBM 1620 to do analyses of variance (a statistical test) on data that had been collected by studying a program for early childhood education, a forerunner to Head Start. Hence there was considerable drama involved in the effort, as the fate of this national educational initiative rode on our work. The algorithms and data being analyzed were sufficiently complex that we were not able to anticipate what answers the computer would come up with. The answers were, of course, determined by the data, but they were not predictable. It turns out that the distinction between being *determined* and being *predictable* is an important one, to which I will return.

I remember how exciting it was when the front-panel lights dimmed right before the algorithm finished its deliberations, as if the computer were deep in thought. When people came by, eager to get the next set of results, I would point to the gently flashing lights and say, "It's thinking." This both was and wasn't a joke—it really *did* seem to be contemplating the answers—and staff members started to ascribe a personality to the machine. It was an anthropomorphization, perhaps, but it did get me to begin to consider in earnest the relationship between thinking and computing.

In order to assess the extent to which my own brain is similar to the computer programs I was familiar with, I began to think about what my brain must be doing as it processed information. I have continued this investigation for fifty years. What I will describe below about our current understanding of how the brain works will sound very different from the standard concept of a computer. Fundamentally, however, the brain does store and process information, and because of the universality of computation—a concept to which I will also return—there is more of a parallel between brains and computers than may be apparent.

Each time I do something—or think of something—whether it is brushing my teeth, walking across the kitchen, contemplating a business problem, practicing on a music keyboard, or coming up with a new idea, I reflect on how I was able to accomplish it. I think even more about all of the things that I am not able to do, as the limitations of human thought provide an equally important set of clues. Thinking so much about thinking might very well be slowing me down, but I have been hopeful that such exercises in self-reflection will enable me to refine my mental methods.

To raise our own awareness of how our brains work, let's consider a series of mind experiments.

Try this: *Recite the alphabet.*

You probably remember this from childhood and can do it easily.

Okay, now try this: *Recite the alphabet backward.*

Unless you have studied the alphabet in this order, you are likely to find it impossible to do. On occasion someone who has spent a significant amount of time in an elementary school classroom where the alphabet is displayed will be able to call up his visual memory and then read it backward from that. Even this is difficult, though, because we do not actually remember whole images. Reciting the alphabet backward should be a simple task, as it involves exactly the same information as reciting it forward, yet we are generally unable to do it.

Do you remember your social security number? If you do, can you recite it backward without first writing it down? How about the nursery rhyme "Mary Had a Little Lamb"? Computers can do this trivially. Yet we fail at it unless we specifically learn the backward sequence as a new series. This tells us something important about how human memory is organized.

Of course, we are able to perform this task easily if we write down the sequence and then read it backward. In doing so we are using a technology—written language—to compensate for one of the limitations of our unaided thinking, albeit a very early tool. (It was our second invention, with spoken language as the first.) This is why we invent tools—to compensate for our shortcomings.

This suggests that **our memories are sequential and in order. They can be accessed in the order that they are remembered. We are unable to directly reverse the sequence of a memory.**

We also have some difficulty starting a memory in the middle of a sequence. If I learn to play a piece of music on the piano, I generally can't just begin it at an arbitrary point in its middle. There are a few points at which I can jump in, because my sequential memory of the piece is organized in segments. If I try to start in the middle of a segment, though, I need to revert to sight-reading until my sequential memory kicks in.

Next, try this: *Recall a walk that you took in the last day or so. What do you remember about it?*

This mind experiment works best if you took a walk very recently, such as earlier today or yesterday. (You can also substitute a drive, or basically any activity during which you moved across some terrain.)

It is likely that you don't remember much about the experience. Who was the fifth person you encountered (not just including people you know)? Did you see an oak tree? A mailbox? What did you see when you turned the first corner? If you passed some stores, what was in the second window? Perhaps you can reconstruct the answers to some of these questions from the few clues that you do remember, but it is likely that you remember relatively few details, even though this is a very recent experience.

If you take walks regularly, think back to the first walk you took last month (or to the first trip to the office last month, if you commute). You probably cannot recall the specific walk or commute at all, and if you do, you doubtless recall even fewer details about it than about your walk today.

I will later discuss the issue of consciousness and make the point that we tend to equate consciousness with our memory of events. The primary reason we believe that we are not conscious when under anesthesia is that we don't remember anything from that period (albeit there are intriguing—and disturbing—exceptions to this). So with regard to the walk I took this morning, was I not conscious during most of it? It's a reasonable question, given that I remember almost nothing about what I saw or even what I was thinking about.

There happen to be a few things I do remember from my walk this morning. I recall thinking about this book, but I couldn't tell you exactly what those thoughts were. I also recall passing a woman pushing a baby carriage. I remember that the woman was attractive, and that the baby was cute as well. I recall two thoughts I had in connection with this experience: *This baby is adorable, like my new grandson,* and *What is this baby perceiving in her visual surroundings?* I cannot recall what either of them was wearing or the color of their hair. (My wife will tell you that that is typical.) Although I am unable to describe anything specific about their appearance, I do have some ineffable sense of what the mom looked like and believe I could pick out her picture from among those of several different women. So while there must be something about her appearance that I have retained in my memory, if I think about the woman, baby carriage, and baby, I am unable to visualize them. There is no photograph or video of this event in my mind. It is hard to describe exactly what *is* in my mind about this experience.

I also recall having passed a different woman with a baby carriage on a walk a few weeks earlier. In that case I don't believe I could even recognize that woman's picture. That memory is now much dimmer than it must have been shortly after that walk.

Next, think about people whom you have encountered only once or twice. Can you visualize them clearly? If you are a visual artist, then you may have learned this observational skill, but typically we are unable to visualize people we've only casually come across to draw or describe them sufficiently but would have little difficulty in recognizing a picture of them.

This suggests that **there are no images, videos, or sound recordings stored in the brain. Our memories are stored as sequences of patterns. Memories that are not accessed dim over time.** When police sketch artists interview a crime victim, they do not ask, "What did the perpetrator's eyebrows look like?" Rather, they will show a series of images of eyebrows and ask the victim to select one. The correct set of eyebrows will trigger the recognition of the same pattern that is stored in the victim's memory.

Let's now consider faces that you know well. *Can you recognize any of these people?*

You are undoubtedly able to recognize these familiar personalities, even though they are partially covered or distorted. This represents a key strength of human perception: **We can recognize a pattern even if only part of it is perceived (seen, heard, felt) and even if it contains alterations. Our recognition ability is apparently able to detect invariant features of a pattern—characteristics that survive real-world variations.** The apparent distortions in a caricature or in certain forms of art such as impressionism emphasize the patterns of an image (person, object) that we recognize while changing other details. The world of art is actually ahead of the world of science in appreciating the power of the human perceptual system. We use the same approach when we recognize a melody from only a few notes.

Now consider this image:

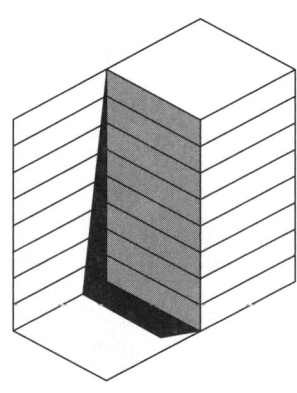

The image is ambiguous—the corner indicated by the black region may be an inside corner or an outside corner. At first you are likely to perceive it one way or the other, though with some effort you can change your perception to the alternate interpretation. Once your mind has fixed on an understanding, however, it may be difficult to see the other perspective. (This turns out to be true of intellectual perspectives as well.) Your brain's interpretation of the image actually influences your experience of it. When the corner appears to be an inside one, your brain will interpret the grey region as a shadow, so it does not seem to be as dark as when you interpret the corner as being an outside one.

Thus **our conscious experience of our perceptions is actually changed by our interpretations.**

Consider that we see what we expect to ___

I'm confident that you were able to complete the above sentence.

Had I written out the last word, you would have needed only to glance at it momentarily to confirm that it was what you had expected.

This implies that **we are constantly predicting the future and hypothesizing what we will experience. This expectation influences what we actually perceive.** Predicting the future is actually the primary reason that we have a brain.

Consider an experience that we all have on a regular basis: A memory from years ago inexplicably pops into your head.

Often this will be a memory of a person or an event that you haven't thought about for a long time. It is evident that something has triggered the memory. The train of thought that did so may be apparent and something you are able to articulate. At other times you may be aware of the sequence of thoughts that led to the memory but would have a hard time expressing it. Often the trigger is quickly lost, so the memory appears to have come from nowhere. I often experience these random memories while doing routine procedures such as brushing my teeth. Sometimes I may be aware of the connection—the toothpaste falling off the toothbrush might remind me of the paint falling off a brush in a painting class I took in college. Sometimes I have only a vague sense of the connection, or none at all.

A related phenomenon that everyone experiences frequently is trying to think of a name or a word. The procedure we use in this circumstance is to try to remind ourselves of triggers that may unlock the memory. (For example: *Who played Queen Padmé in* Revenge of the Sith? *Let's see, it's that same actress who was the star in a recent dark movie about dancing, that was* Black Swan, *oh yes, Natalie Portman.*) Sometimes we adopt idiosyncratic mnemonics to help us remember. (For example: *She's always slim, not portly, oh yes, Portman, Natalie Portman.*) Some of our memories are sufficiently robust that we can go directly from a question (such as *who played Queen Padmé*) to the answer; often we need to go through a series of triggers until we find one that works. It's very much like having the right Web link. Memories can indeed become lost like a Web page to which no other page links to (at least no page that we can find).

While executing routine procedures—such as putting on a shirt—watch yourself performing them, and consider the extent to which you follow the same sequence of steps each time. From my own observation (and as I mentioned, I am constantly trying to observe myself), it is likely that you follow very much the same steps each time you perform a particular routine task, though there may be additional modules added. For example, most of my shirts do not require cuff links, but when one does, that involves a further series of tasks.

The lists of steps in my mind are organized in hierarchies. I follow a routine procedure before going to sleep. The first step is to brush my teeth. But this action is in turn broken into a smaller series of steps, the first of which is to put toothpaste on the toothbrush. That step in turn is made up of yet smaller steps, such as finding the toothpaste, removing the cap, and so on. The step of finding the toothpaste also has steps, the first of which is to open the bathroom cabinet. That step in turn requires steps, the first of which is to grab the outside of the cabinet door. This nesting actually continues down to a very fine grain of movements, so that there are literally thousands of little actions constituting my nighttime routine. Although I may have difficulty remembering details of a walk I took just a few hours ago, I have no difficulty recalling all of these many steps in preparing for bed—so much so that I am able to think about other things while I go through these procedures. It is important to point out that this list is not stored as one long list of thousands of steps—**rather, each of our routine procedures is remembered as an elaborate hierarchy of nested activities.**

The same type of hierarchy is involved in our ability to recognize objects and situations. We recognize the faces of people we know well and also recognize that these faces contain eyes, a nose, a mouth, and so on—a hierarchy of patterns that we use in both our perceptions and our actions. The use of hierarchies allows us to reuse patterns. For example, we do not need to relearn the concept of a nose and a mouth each time we are introduced to a new face.

In the next chapter, we'll put the results of these thought experiments together into a theory of how the neocortex must work. I will argue that they reveal essential attributes of our thinking that are uniform, from finding the toothpaste to writing a poem.

CHAPTER 3

A MODEL OF THE NEOCORTEX: THE PATTERN RECOGNITION THEORY OF MIND

The brain is a tissue. It is a complicated, intricately woven tissue, like nothing else we know of in the universe, but it is composed of cells, as any tissue is. They are, to be sure, highly specialized cells, but they function according to the laws that govern any other cells. Their electrical and chemical signals can be detected, recorded and interpreted and their chemicals can be identified; the connections that constitute the brain's woven feltwork can be mapped. In short, the brain can be studied, just as the kidney can.

—David H. Hubel, neuroscientist

Suppose that there be a machine, the structure of which produces thinking, feeling, and perceiving; imagine this machine enlarged but preserving the same proportions, so you could enter it as if it were a mill. This being supposed, you might visit inside; but what would you observe there? Nothing but parts which push and move each other, and never anything that could explain perception.

—Gottfried Wilhelm Leibniz

A Hierarchy of Patterns

I have repeated the simple experiments and observations described in the previous chapter thousands of times in myriad contexts. The conclusions from these observations necessarily constrain my explanation for what the brain must be doing, just as the simple experiments on time, space, and mass that were conducted in the early and late nineteenth century necessarily constrained the young Master Einstein's reflections on how the universe functioned. In the discussion that follows I'll also factor in some very basic observations from neuroscience, attempting to avoid the many details that are still in contention.

First, let me explain why this section specifically discusses the neocortex (from the Latin meaning "new rind"). We do know the neocortex is responsible for our ability to deal with patterns of information and to do so in a hierarchical fashion. Animals without a neocortex (basically nonmammals) are largely incapable of understanding hierarchies.[1] Understanding and leveraging the innately hierarchical nature of reality is a uniquely mammalian trait and results from mammals' unique possession of this evolutionarily recent brain structure. The neocortex is responsible for sensory perception, recognition of everything from visual objects to abstract concepts, controlling movement, reasoning from spatial orientation to rational thought, and language—basically, what we regard as "thinking."

The human neocortex, the outermost layer of the brain, is a thin, essentially two-dimensional structure with a thickness of about 2.5 millimeters (about a tenth of an inch). In rodents, it is about the size of a postage stamp and is smooth. An evolutionary innovation in primates is that it became intricately folded over the top of the rest of the brain with deep ridges, grooves, and wrinkles to increase its surface area. Due to its elaborate folding, the neocortex constitutes the bulk of the human brain, accounting for 80 percent of its weight. *Homo sapiens* developed a large forehead to allow for an even larger neocortex; in particular we have a frontal lobe where we deal with the more abstract patterns associated with high-level concepts.

This thin structure is basically made up of six layers, numbered I (the outermost layer) to VI. The axons emerging from the neurons in layers II and III project to other parts of the neocortex. The axons (output connections) from layers V and VI are connected primarily outside of the neocortex to the thalamus, brain stem, and spinal cord. The neurons in layer IV receive synaptic (input) connections from neurons that are outside the neocortex, especially in the thalamus. The number of layers varies slightly from region to region. Layer IV is very thin in the motor cortex, because in that area it largely does not receive input from the thalamus, brain stem, or spinal cord. Conversely, in the occipital lobe (the part of the neocortex usually responsible for visual processing), there are three additional sublayers that can be seen in layer IV, due to the considerable input flowing into this region, including from the thalamus.

A critically important observation about the neocortex is the extraordinary uniformity of its fundamental structure. This was first noticed by American neuroscientist Vernon Mountcastle (born in 1918). In 1957 Mountcastle discovered the columnar organization of the neocortex. In 1978 he made an observation that is as significant to neuroscience as the Michelson-Morley ether-disproving experiment of 1887 were to physics. That year he described the remarkably unvarying organization of the neocortex, hypothesizing that it was composed of a single mechanism that was repeated over and over again,[2] and proposing the cortical column as that basic unit. The differences in the height of certain layers in different regions noted above are simply differences in the amount of interconnectivity that the regions are responsible for dealing with.

Mountcastle hypothesized the existence of mini-columns within columns, but this theory became controversial because there were no visible demarcations of such smaller structures. However, extensive experimentation has revealed that there are in fact repeating units within the neuron fabric of each column. It is my contention that the basic unit is a pattern recognizer and that this constitutes the fundamental component of the neocortex. In contrast to Mountcastle's notion of a mini-column, there is no specific physical boundary to these recognizers, as they are placed closely one to the next in an interwoven fashion, so the cortical column is simply an aggregate of a large number of them. These recognizers are capable of wiring themselves to one another throughout the course of a lifetime, so the elaborate connectivity (between modules) that we see in the neocortex is not prespecified by the genetic code, but rather is created to reflect the patterns we actually learn over time. I will describe this thesis in more detail, but I maintain that this is how the neocortex must be organized.

It should be noted, before we further consider the structure of the neocortex, that it is important to model systems at the right level. Although chemistry is theoretically based on physics and could be derived entirely from physics, this would be unwieldy and infeasible in practice, so chemistry has established its own rules and models. Similarly, we should be able to deduce the laws of thermodynamics from physics, but once we have a sufficient number of particles to call them a gas rather than simply a bunch of particles, solving equations for the physics of each particle interaction becomes hopeless, whereas the laws of thermodynamics work quite well. Biology likewise has its own rules and models. A single pancreatic islet cell is enormously complicated, especially if we model it at the level of molecules; modeling what a pancreas actually does in terms of regulating levels of insulin and digestive enzymes is considerably less complex.

The same principle applies to the levels of modeling and understanding in the brain. It is certainly a useful and necessary part of reverse-engineering the brain to model its interactions at the molecular level, but the goal of the effort here is essentially to refine our model to account for how the brain processes information to produce cognitive meaning.

American scientist Herbert A. Simon (1916–2001), who is credited with cofounding the field of artificial intelligence, wrote eloquently about the issue of understanding complex systems at the right level of abstraction. In describing an AI program he had devised called EPAM (elementary perceiver and memorizer), he wrote in 1973, "Suppose you decided that you wanted to understand the mysterious EPAM program that I have. I could provide you with two versions of it. One would be . . . the form in which it was actually written—with its whole structure of routines and subroutines. . . . Alternatively, I could provide you with a machine-language version of EPAM after the whole translation had been carried out—after it had been flattened so to speak. . . . I don't think I need argue at length which of these two versions would provide the most parsimonious, the most meaningful, the most lawful description. . . . I will not even propose to you the third . . . of providing you with neither program, but instead with the electromagnetic equations and boundary conditions that the computer, viewed as a physical system, would have to obey while behaving as EPAM. That would be the acme of reduction and incomprehensibility."[3]

There are about a half million cortical columns in a human neocortex, each occupying a space about two millimeters high and a half millimeter wide and containing about 60,000 neurons (resulting in a total of about 30 billion neurons in the neocortex). A rough estimate is that each pattern recognizer within a cortical column contains about 100 neurons, so there are on the order of 300 million pattern recognizers in total in the neocortex.

As we consider how these pattern recognizers work, let me begin by saying that it is difficult to know precisely where to begin. Everything happens simultaneously in the neocortex, so there is no beginning and no end to its processes. I will frequently need to refer to phenomena that I have not yet explained but plan to come back to, so please bear with these forward references.

Human beings have only a weak ability to process logic, but a very deep core capability of recognizing patterns. To do logical thinking, we need to use the neocortex, which is basically a large pattern recognizer. It is not an ideal mechanism for performing logical transformations, but it is the only facility we have for the job. Compare, for example, how a human plays chess to how a typical computer chess program works. Deep Blue, the computer that defeated Garry Kasparov, the human world chess champion, in 1997 was capable of analyzing the logical implications of 200 million board positions (representing different move-countermove sequences) every second. (That can now be done, by the way, on a few personal computers.) Kasparov was asked how many positions he could analyze each second, and he said it was less than one. How is it, then, that he was able to hold up to Deep Blue at all? The answer is the very strong ability humans have to recognize patterns. However, we need to train this facility, which is why not everyone can play master chess.

Kasparov had learned about 100,000 board positions. That's a real number—we have established that a human master in a particular field has mastered about 100,000 chunks of knowledge. Shakespeare composed his plays with 100,000 word senses (employing about 29,000 distinct words, but using most of them in multiple ways). Medical expert systems that have been built to represent the knowledge of a human medical physician have shown that a typical human medical specialist has mastered about 100,000 concepts in his or her domain. Recognizing a chunk of knowledge from this store is not straightforward, as a particular item will present itself a little bit differently each time it is experienced.

Armed with his knowledge, Kasparov looks at the chessboard and compares the patterns that he sees to all 100,000 board situations that he has mastered, and he does all 100,000 comparisons simultaneously. There is consensus on this point: All of our neurons are processing—considering the patterns—at the same time. That does not mean that they are all *firing* simultaneously (we would probably fall to the floor if that happened), but while doing their processing are considering the possibility of firing.

How many patterns can the neocortex store? We need to factor in the phenomenon of redundancy. The face of a loved one, for example, is not stored once but on the order of thousands of times. Some of these repetitions are largely the same image of the face, whereas most show different perspectives of it, different lighting, different expressions, and so on. None of these repeated patterns are stored as images per se (that is, as two-dimensional arrays of pixels). Rather, they are stored as lists of features where the constituent elements of a pattern are themselves patterns. We'll describe below more precisely what these hierarchies of features look like and how they are organized.

If we take the core knowledge of an expert as consisting of about 100,000 "chunks" of knowledge (that is, patterns) with a redundancy estimate of about 100 to 1, that gives us a requirement of 10 million patterns. This core expert knowledge is built on more general and extensive professional knowledge, so we can increase the order of magnitude of patterns to about 30 to 50 million. Our everyday "commonsense" knowledge as a human being is even greater; "street smarts" actually require substantially more of our neocortex than "book smarts." Including this brings our estimate to well over 100 million patterns, taking into account the redundancy factor of about 100. Note that the redundancy factor is far from fixed—very common patterns will have a redundancy factor well into the thousands, whereas a brand-new phenomenon may have a redundancy factor of less than 10.

As I will discuss below, our procedures and actions also comprise patterns and are likewise stored in regions of the cortex, so my estimate of the total capacity of the human neocortex is on the order of low hundreds of millions of patterns. This rough tally correlates well with the number of pattern recognizers that I estimated above at about 300 million, so it is a reasonable conclusion that the function of each neocortical pattern recognizer is to process one iteration (that is, one copy among the multiple redundant copies of most patterns in the neocortex) of a pattern. Our estimates of the number of patterns that a human brain is capable of dealing with (including necessary redundancy) and the number of physical pattern recognizers happen to be the same order of magnitude. It should be noted here that when I refer to "processing" a pattern, I am referring to all of the things we are able to do with a pattern: learn it, predict it (including parts of it), recognize it, and implement it (either by thinking about it further or through a pattern of physical movement).

Three hundred million pattern processors may sound like a large number, and indeed it was sufficient to enable *Homo sapiens* to develop verbal and written language, all of our tools, and other diverse creations. These inventions have built upon themselves, giving rise to the exponential growth of the information content of technologies as described in my law of accelerating returns. No other species has achieved this. As I discussed, a few other species, such as chimpanzees, do appear to have a rudimentary ability to understand and form language and also to use primitive tools. They do, after all, also have a neocortex, but their abilities are limited due to its smaller size, especially of the frontal lobe. The size of our own neocortex has exceeded a threshold that has enabled our species to build ever more powerful tools, including tools that can now enable us to understand our own intelligence. Ultimately our brains, combined with the technologies they have fostered, will permit us to create a synthetic neocortex that will contain well beyond a mere 300 million pattern processors. Why not a billion? Or a trillion?

The Structure of a Pattern

The pattern recognition theory of mind that I present here is based on the recognition of patterns by pattern recognition modules in the neocortex. These patterns (and the modules) are organized in hierarchies. I discuss below the intellectual roots of this idea, including my own work with hierarchical pattern recognition in the 1980s and 1990s and Jeff Hawkins (born in 1957) and Dileep George's (born in 1977) model of the neocortex in the early 2000s.

Each pattern (which is recognized by one of the estimated 300 million pattern recognizers in the neocortex) is composed of three parts. Part one is the input, which consists of the lower-level patterns that compose the main pattern. The descriptions for each of these lower-level patterns do not

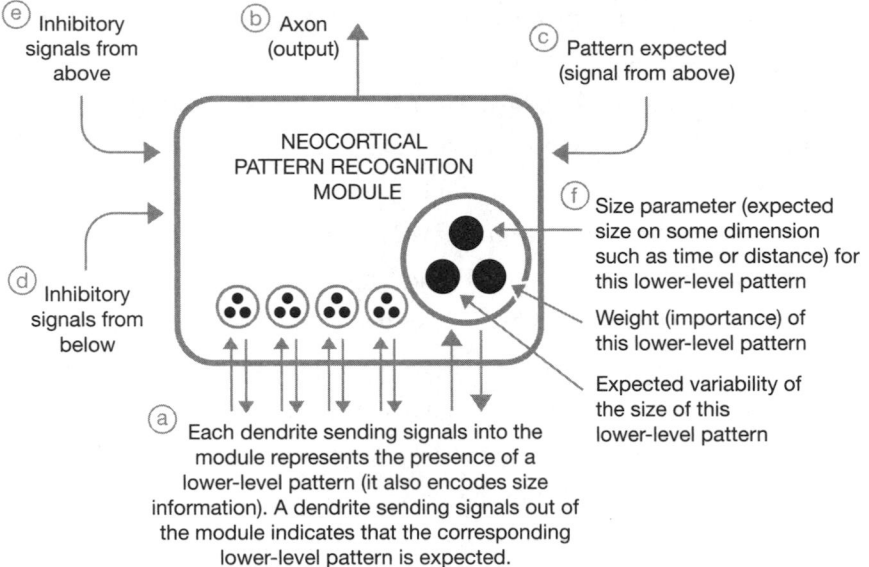

(e) Inhibitory signals from above

(b) Axon (output)

(c) Pattern expected (signal from above)

NEOCORTICAL PATTERN RECOGNITION MODULE

(d) Inhibitory signals from below

(f) Size parameter (expected size on some dimension such as time or distance) for this lower-level pattern

Weight (importance) of this lower-level pattern

Expected variability of the size of this lower-level pattern

(a) Each dendrite sending signals into the module represents the presence of a lower-level pattern (it also encodes size information). A dendrite sending signals out of the module indicates that the corresponding lower-level pattern is expected.

need to be repeated for each higher-level pattern that references them. For example, many of the patterns for words will include the letter "A." Each of these patterns does not need to repeat the description of the letter "A" but will use the same description. Think of it as being like a Web pointer. There is one Web page (that is, one pattern) for the letter "A," and all of the Web pages (patterns) for words that include "A" will have a link to the "A" page (to the "A" pattern). Instead of Web links, the neocortex uses actual neural connections. There is an axon from the "A" pattern recognizer that connects to multiple dendrites, one for each word that uses "A." Keep in mind also the redundancy factor: There is more than one pattern recognizer for the letter "A." Any of these multiple "A" pattern recognizers can send a signal up to the pattern recognizers that incorporate "A."

The second part of each pattern is the pattern's name. In the world of language, this higher-level pattern is simply the word "apple." Although we directly use our neocortex to understand and process every level of language, most of the patterns it contains are not language patterns per se. In the neocortex the "name" of a pattern is simply the axon that emerges from

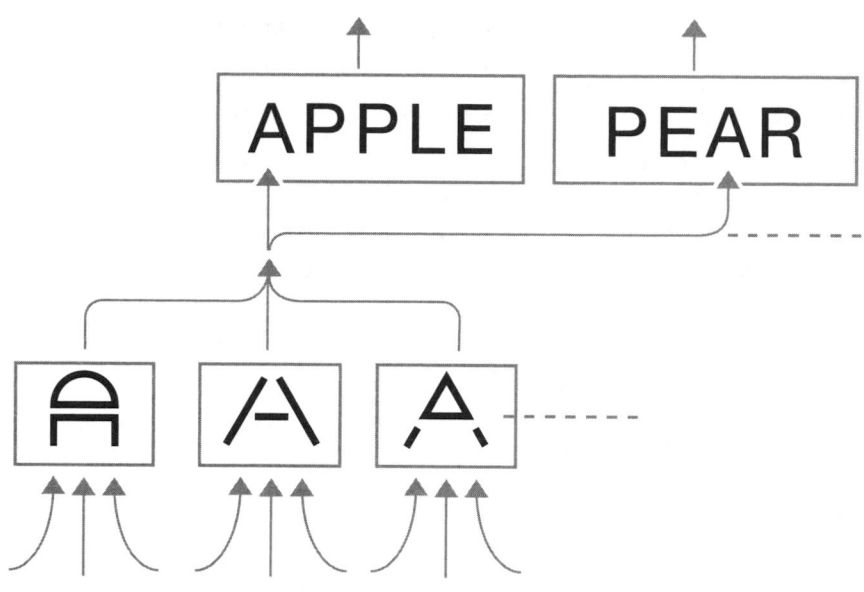

Three redundant (but somewhat different) patterns for "A" feeding up to higher-level patterns that incorporate "A."

each pattern processor; when that axon fires, its corresponding pattern has been recognized. The firing of the axon is that pattern recognizer shouting the name of the pattern: "Hey guys, I just saw the written word 'apple.'"

The third and final part of each pattern is the set of higher-level patterns that it in turn is part of. For the letter "A," this is all of the words that include "A." These are, again, like Web links. Each recognized pattern at one level triggers the next level that part of that higher-level pattern is present. In the neocortex, these links are represented by physical dendrites that flow into neurons in each cortical pattern recognizer. Keep in mind that each neuron can receive inputs from multiple dendrites yet produces a single output on an axon. That axon, however, can then in turn transmit to multiple dendrites.

To take some simple examples, the simple patterns on the next page are a small subset of the patterns used to make up printed letters. Note that every level constitutes a pattern. In this case, the shapes are patterns, the letters are patterns, and the words are also patterns. Each of these patterns has a set of inputs, a process of pattern recognition (based on the inputs that take place in the module), and an output (which feeds to the next higher level of pattern recognizer).

Southwest to north-central connection:

Southeast to north-central connection:

Horizontal crossbar:

Leftmost vertical line:

Concave region facing south:

Bottom horizontal line:

Top horizontal line:

Middle horizontal line:

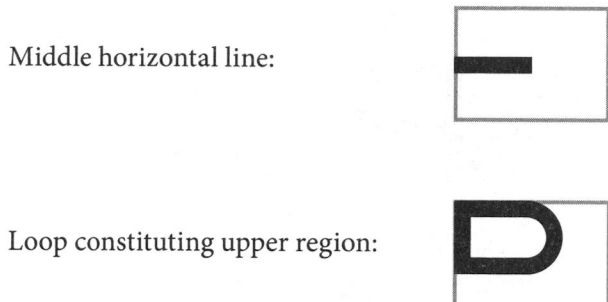

Loop constituting upper region:

The above patterns are constituents of the next higher level of pattern, which is a category called printed letters (there is no such formal category within the neocortex, however; indeed, there are no formal categories).

"A":

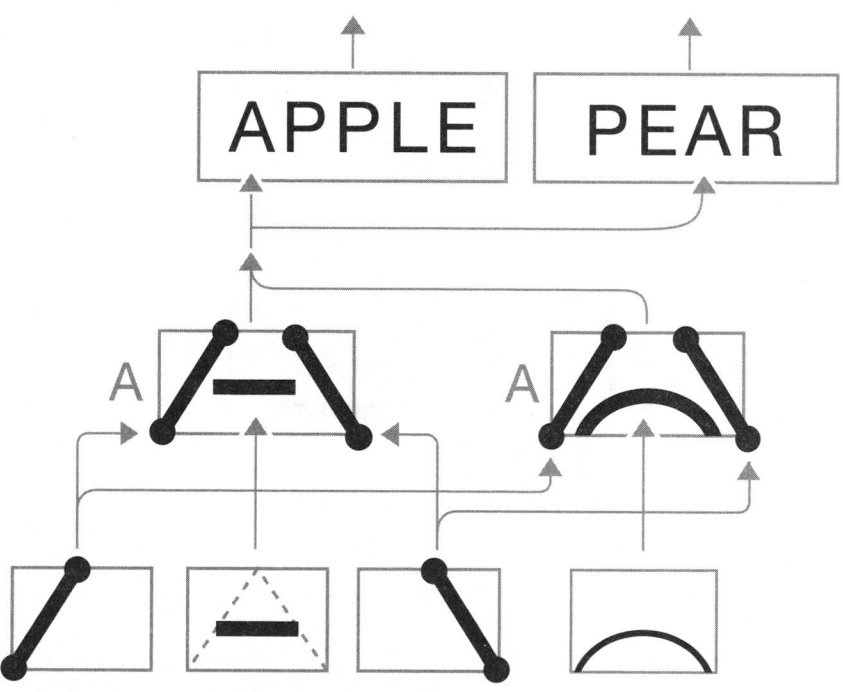

Two different patterns, either of which constitutes "A," and two different patterns at a higher level ("APPLE" and "PEAR") of which "A" is a part.

"P":

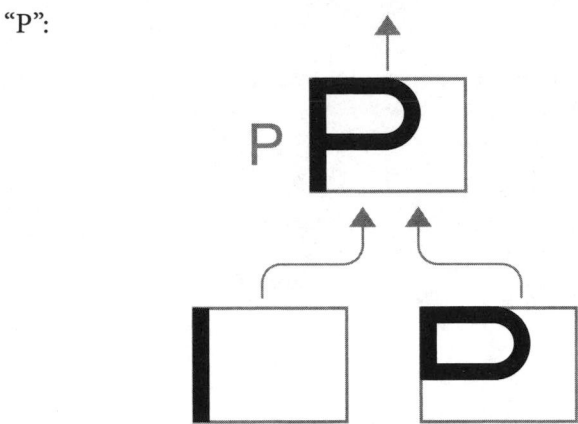

Patterns that are part of the higher-level pattern "P."

"L":

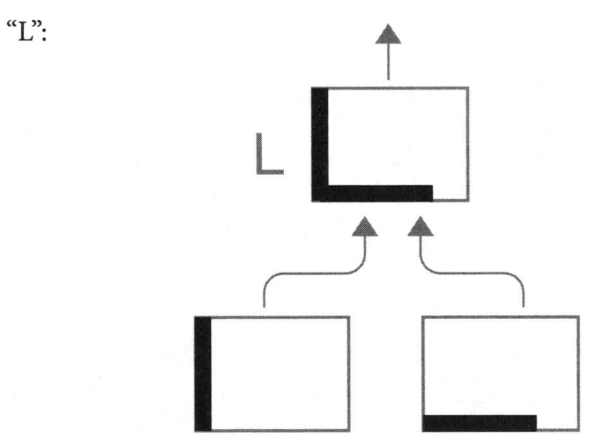

Patterns that are part of the higher-level pattern "L."

"E":

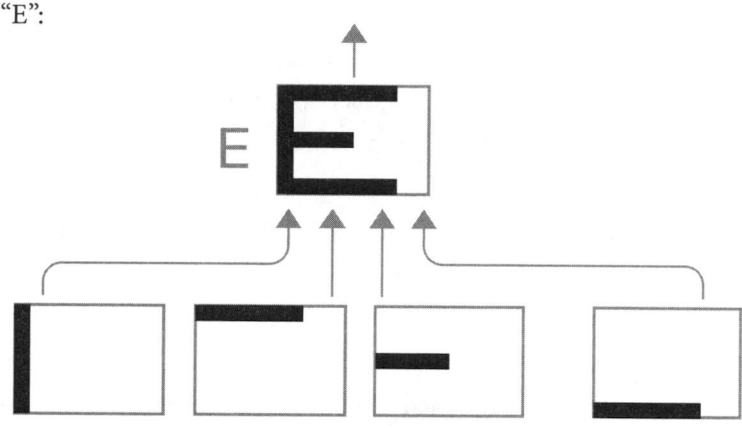

Patterns that are part of the higher-level pattern "E."

These letter patterns feed up to an even higher-level pattern in a category called words. (The word "words" is our language category for this concept, but the neocortex just treats them only as patterns.)

"APPLE":

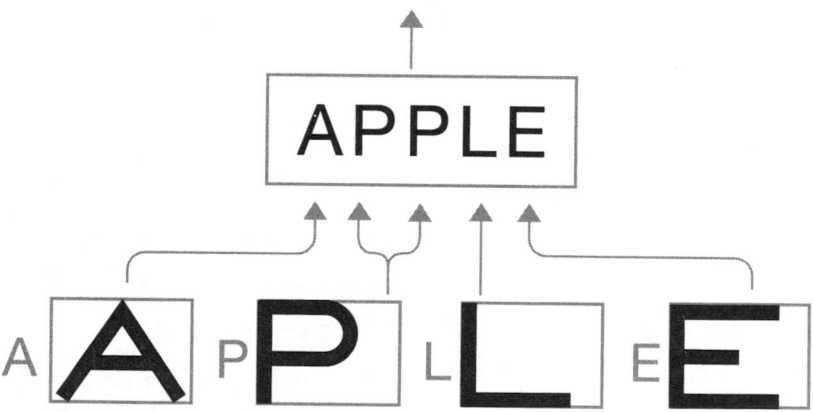

In a different part of the cortex is a comparable hierarchy of pattern recognizers processing actual *images* of objects (as opposed to printed letters). If you are looking at an actual apple, low-level recognizers will detect curved edges and surface color patterns leading up to a pattern recognizer firing its axon and saying in effect, "Hey guys, I just saw an actual apple." Yet other pattern recognizers will detect combinations of frequencies of sound leading up to a pattern recognizer in the auditory cortex that might fire its axon indicating, "I just heard the spoken word 'apple.'"

Keep in mind the redundancy factor—we don't just have a single pattern recognizer for "apple" in each of its forms (written, spoken, visual). There are likely to be hundreds of such recognizers firing, if not more. The redundancy not only increases the likelihood that you will successfully recognize each instance of an apple but also deals with the variations in real-world apples. For apple objects, there will be pattern recognizers that deal with the many varied forms of apples: different views, colors, shadings, shapes, and varieties.

Also keep in mind that the hierarchy shown above is a hierarchy of *concepts*. These recognizers are not physically placed above each other; because of the thin construction of the neocortex, it is physically only one pattern recognizer high. The conceptual hierarchy is created by the connections between the individual pattern recognizers.

An important attribute of the PRTM is how the recognitions are made inside each pattern recognition module. Stored in the module is a weight for each input dendrite indicating how important that input is to the recognition. The pattern recognizer has a threshold for firing (which indicates that this pattern recognizer has successfully recognized the pattern it is responsible for). Not every input pattern has to be present for a recognizer to fire. The recognizer may still fire if an input with a low weight is missing, but it is less likely to fire if a high-importance input is missing. When it fires, a pattern recognizer is basically saying, "The pattern I am responsible for is probably present."

Successful recognition by a module of its pattern goes beyond just counting the input signals that are activated (even a count weighted by the importance parameter). The size (of each input) matters. There is another parameter (for each input) indicating the expected size of the input, and yet another indicating how variable that size is. To appreciate how this works, suppose we have a pattern recognizer that is responsible for recognizing the spoken word "steep." This spoken word has four sounds: [s], [t], [E], and [p]. The [t] phoneme is what is known as a "dental consonant," meaning that it is created by the tongue creating a burst of noise when air breaks its contact with the upper teeth. It is essentially impossible to articulate the [t] phoneme slowly. The [p] phoneme is considered a "plosive consonant" or "oral occlusive," meaning that it is created when the vocal tract is suddenly blocked (by the lips in the case of [p]) so that air no longer passes. It is also necessarily quick. The [E] vowel is caused by resonances of the vocal cord and open mouth. It is considered a "long vowel," meaning that it persists for a much longer period of time than consonants such as [t] and [p]; however, its duration can be quite variable. The [s] phoneme is known as a "sibilant consonant," and is caused by the passage of air against the edges of the teeth, which are held close together. Its duration is typically shorter than that of a long vowel such as [E], but it is also variable (in other words, the [s] can be said quickly or you can drag it out).

In our work in speech recognition, we found that it is necessary to encode this type of information in order to recognize speech patterns. For example, the words "step" and "steep" are very similar. Although the [e] phoneme in "step" and the [E] in "steep" are somewhat different vowel sounds (in that they have different resonant frequencies), it is not reliable to distinguish these two words based on these often confusable vowel sounds. It is much more reliable to consider the observation that the [e] in "step" is relatively brief compared with the [E] in "steep."

We can encode this type of information with two numbers for each input: the expected size and the degree of variability of that size. In our "steep" example, [t] and [p] would both have a very short expected duration as well as a small expected variability (that is, we do not expect to hear long t's and p's). The [s] sound would have a short expected duration but a larger variability because it is possible to drag it out. The [E] sound has a long expected duration as well as a high degree of variability.

In our speech examples, the "size" parameter refers to duration, but time is only one possible dimension. In our work in character recognition, we found that comparable spatial information was important in order to recognize printed letters (for example the dot over the letter "i" is expected to be much smaller than the portion under the dot). At much higher levels of abstraction, the neocortex will deal with patterns with all sorts of continuums, such as levels of attractiveness, irony, happiness, frustration, and myriad others. We can draw similarities across rather diverse continuums, as Darwin did when he related the physical size of geological canyons to the amount of differentiation among species.

In a biological brain, the source of these parameters comes from the brain's own experience. We are not born with an innate knowledge of phonemes; indeed different languages have very different sets of them. This implies that multiple examples of a pattern are encoded in the learned parameters of each pattern recognizer (as it requires multiple instances of a pattern to ascertain the expected distribution of magnitudes of the inputs to the pattern). In some AI systems, these types of parameters are hand-coded by experts (for example, linguists who can tell us the expected durations of different phonemes, as I articulated above). In my own work, we found that having an AI system discover these parameters on its own from training data (similar to the way the brain does it) was a superior approach. Sometimes we used a hybrid approach; that is, we primed the system with the intuition of human experts (for the initial settings of the parameters) and then had the AI system automatically refine these estimates using a learning process from real examples of speech.

What the pattern recognition module is doing is computing the probability (that is, the likelihood based on all of its previous experience) that the pattern that it is responsible for recognizing is in fact currently represented by its active inputs. Each particular input to the module is active if the corresponding lower-level pattern recognizer is firing (meaning that that lower-level pattern was recognized). Each input also encodes the observed size (on some appropriate dimension such as temporal duration or physical magnitude or some other continuum) so that the size can be compared (with the stored size parameters for each input) by the module in computing the overall probability of the pattern.

How does the brain (and how can an AI system) compute the overall probability that the pattern (that the module is responsible for recognizing) is present given (1) the inputs (each with an observed size), (2) the stored parameters on size (the expected size and the variability of size) for each input, and (3) the parameters of the importance of each input? In the 1980s and 1990s, I and others pioneered a mathematical method called hierarchical hidden Markov models for learning these parameters and then using them to recognize hierarchical patterns. We used this technique in the recognition of human speech as well as the understanding of natural language. I describe this approach further in chapter 7.

Getting back to the flow of recognition from one level of pattern recognizers to the next, in the above example we see the information flow up the conceptual hierarchy from basic letter features to letters to words. Recognitions will continue to flow up from there to phrases and then more complex language structures. If we go up several dozen more levels, we get to higher-level concepts like irony and envy. Even though every pattern recognizer is working simultaneously, it does take time for recognitions to move upward in this conceptual hierarchy. Traversing each level takes between a few hundredths to a few tenths of a second to process. Experiments have shown that a moderately high-level pattern such as a face takes at least a tenth of a second. It can take as long as an entire second if there are significant distortions. If the brain were sequential (like conventional computers) and was performing each pattern recognition in sequence, it would have to consider every possible low-level pattern before moving on to the next level. Thus it would take many millions of cycles just to go through each level. That is exactly what happens when we simulate these processes on a computer. Keep in mind, however, that computers process millions of times faster than our biological circuits.

A very important point to note here is that information flows down the conceptual hierarchy as well as up. If anything, this downward flow is even more significant. If, for example, we are reading from left to right and have already seen and recognized the letters "A," "P," "P," and "L," the "APPLE" recognizer will predict that it is likely to see an "E" in the next position. It will send a signal *down* to the "E" recognizer saying, in effect, "Please be aware that there is a high likelihood that you will see your 'E' pattern very soon, so be on the lookout for it." The "E" recognizer then adjusts its threshold such that it is more likely to recognize an "E." So if an image appears next that is vaguely like an "E," but is perhaps smudged such that it would not have been recognized as an "E" under "normal" circumstances, the "E" recognizer may nonetheless indicate that it has indeed seen an "E," since it was expected.

The neocortex is, therefore, predicting what it expects to encounter. Envisaging the future is one of the primary reasons we have a neocortex. At the highest conceptual level, we are continually making predictions— who is going to walk through the door next, what someone is likely to say next, what we expect to see when we turn the corner, the likely results of our own actions, and so on. These predictions are constantly occurring at *every* level of the neocortex hierarchy. We often misrecognize people and things and words because our threshold for confirming an expected pattern is too low.

In addition to positive signals, there are also negative or inhibitory signals which indicate that a certain pattern is less likely to exist. These can come from lower conceptual levels (for example, the recognition of a mustache will inhibit the likelihood that a person I see in the checkout line is my wife), or from a higher level (for example, I know that my wife is on a trip, so the person in the checkout line can't be she). When a pattern recognizer receives an inhibitory signal, it raises the recognition threshold, but it is still possible for the pattern to fire (so if the person in line really is her, I may still recognize her).

The Nature of the Data Flowing into a Neocortical Pattern Recognizer

Let's consider further what the data for a pattern looks like. If the pattern is a face, the data exists in at least two dimensions. We cannot say that the eyes necessarily come first, followed by the nose, and so on. The same thing is true for most sounds. A musical piece has at least two dimensions. There may be more than one instrument and/or voice making sounds at the same time. Moreover, a single note of a complex instrument such as the piano consists of multiple frequencies. A single human voice consists of varying levels of energy in dozens of different frequency bands simultaneously. So a pattern of sound may be complex at any one instant, and these complex instants stretch out over time. Tactile inputs are also two-dimensional, since the skin is a two-dimensional sense organ, and such patterns may change over the third dimension of time.

So it would seem that the input to a neocortex pattern processor must comprise two- if not three-dimensional patterns. However, we can see in the structure of the neocortex that the pattern inputs are only one-dimensional lists. All of our work in the field of creating artificial pattern recognition systems (such as speech recognition and visual recognition systems) demonstrates that we can (and did) represent two- and three-dimensional phenomena with such one-dimensional lists. I'll describe how these methods work in chapter 7, but for now we can proceed with the understanding that the input to each pattern processor is a one-dimensional list, even though the pattern itself may inherently reflect more than one dimension.

We should factor in at this point the insight that the patterns we have learned to recognize (for example, a specific dog or the general idea of a "dog," a musical note or a piece of music) are exactly the same mechanism that is the basis for our memories. Our memories are in fact patterns organized as lists (where each item in each list is another pattern in the cortical hierarchy) that we have learned and then recognize when presented with the appropriate stimulus. In fact, memories exist in the neocortex in order to be recognized.

The only exception to this is at the lowest possible conceptual level, in which the input data to a pattern represents specific sensory information (for example, image data from the optic nerve). Even this lowest level of pattern, however, has been significantly transformed into simple patterns by the time it reaches the cortex. The lists of patterns that constitute a memory are in forward order, and we are able to remember our memories only in that order, hence the difficulty we have in reversing our memories.

A memory needs to be triggered by another thought/memory (these are the same thing). We can experience this mechanism of triggering when we are perceiving a pattern. When we perceived "A," "P," "P," and "L," the "A P P L E" pattern predicted that we would see an "E" and triggered the "E" pattern that it is now expected. Our cortex is thereby "thinking" of seeing an "E" even before we see it. If this particular interaction in our cortex has our attention, we will think about "E" before we see it or even if we never see it. A similar mechanism triggers old memories. Usually there is an entire chain of such links. Even if we do have some level of awareness of the memories (that is, the patterns) that triggered the old memory, memories (patterns) do not have language or image labels. This is the reason why old memories may seem to suddenly jump into our awareness. Having been buried and not activated for perhaps years, they need a trigger in the same way that a Web page needs a Web link to be activated. And just as a Web page can become "orphaned" because no other page links to it, the same thing can happen to our memories.

Our thoughts are largely activated in one of two modes, undirected and directed, both of which use these same cortical links. In the undirected mode, we let the links play themselves out without attempting to move them in any particular direction. Some forms of meditation (such as Transcendental Meditation, which I practice) are based on letting the mind do exactly this. Dreams have this quality as well.

In directed thinking we attempt to step through a more orderly process of recalling a memory (a story, for example) or solving a problem. This also involves stepping through lists in our neocortex, but the less structured flurry of undirected thought will also accompany the process. The full content of our thinking is therefore very disorderly, a phenomenon that James Joyce illuminated in his "stream of consciousness" novels.

As you think through the memories/stories/patterns in your life, whether they involve a chance encounter with a mother with a baby carriage and baby on a walk or the more important narrative of how you met your spouse, your memories consist of a sequence of patterns. Because these patterns are not labeled with words or sounds or pictures or videos, when you try to recall a significant event, you will essentially be reconstructing the images in your mind, because the actual images do not exist.

If we were to "read" the mind of someone and peer at exactly what is going on in her neocortex, it would be very difficult to interpret her memories, whether we were to take a look at patterns that are simply stored in the neocortex waiting to be triggered or those that have been triggered and are currently being experienced as active thoughts. What we would "see" is the simultaneous activation of millions of pattern recognizers. A hundredth of a second later, we would see a different set of a comparable number of activated pattern recognizers. Each such pattern would be a list of other patterns, and each of those patterns would be a list of other patterns, and so on until we reached the most elementary simple patterns at the lowest level. It would be extremely difficult to interpret what these higher-level patterns meant without actually copying *all* of the information at every level into our own cortex. Thus each pattern in our neocortex is meaningful only in light of all the information carried in the levels below it. Moreover, other patterns at the same level and at higher levels are also relevant in interpreting a particular pattern because they provide context. True mind reading, therefore, would necessitate not just detecting the activations of the relevant axons in a person's brain, but examining essentially her entire neocortex with all of its memories to understand these activations.

As we experience our own thoughts and memories, we "know" what they mean, but they do not exist as readily explainable thoughts and recollections. If we want to share them with others, we need to translate them into language. This task is also accomplished by the neocortex, using pattern recognizers trained with patterns that we have learned for the purpose of using language. Language is itself highly hierarchical and evolved to take advantage of the hierarchical nature of the neocortex, which in turn reflects the hierarchical nature of reality. The innate ability of humans to learn the hierarchical structures in language that Noam Chomsky wrote about reflects the structure of the neocortex. In a 2002 paper he coauthored, Chomsky cites the attribute of "recursion" as accounting for the unique language faculty of the human species.[4] Recursion, according to Chomsky, is the ability to put together small parts into a larger chunk, and then use that chunk as a part in yet another structure, and to continue this process iteratively. In this way we are able to build the elaborate structures of sentences and paragraphs from a limited set of words. Although Chomsky was not explicitly referring here to brain structure, the capability he is describing is exactly what the neocortex does.

Lower species of mammals largely use up their neocortex with the challenges of their particular lifestyles. The human species acquired additional capacities by having grown substantially more cortex to handle spoken and written language. Some people have learned such skills better than others. If we have told a particular story many times, we will begin to actually learn the sequence of language that describes the story as a series of separate sequences. Even in this case our memory is not a strict sequence of words, but rather of language structures that we need to translate into specific word sequences each time we deliver the story. That is why we tell a story a bit differently each time we share it (unless we learn the exact word sequence as a pattern).

For each of these descriptions of specific thought processes, we also need to consider the issue of redundancy. As I mentioned, we don't have a single pattern representing the important entities in our lives, whether those entities constitute sensory categories, language concepts, or memories of events. Every important pattern—at every level—is repeated many times. Some of these recurrences represent simple repetitions, whereas many represent different perspectives and vantage points. This is a principal reason why we can recognize a familiar face from various orientations and under a range of lighting conditions. Each level up the hierarchy has substantial redundancy, allowing sufficient variability that is consistent with that concept.

So if we were to imagine examining your neocortex when you were looking at a particular loved one, we would see a great many firings of the axons of the pattern recognizers at every level, from the basic level of

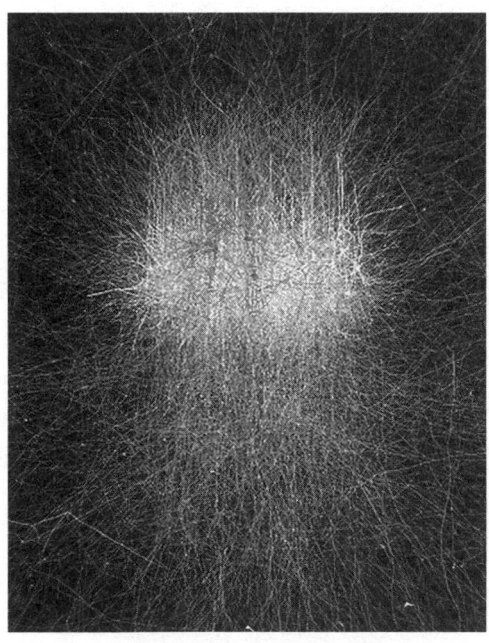

- -
A computer simulation of the firings of many simultaneous pattern recognizers in the neocortex.

primitive sensory patterns up to many different patterns representing that loved one's image. We would also see massive numbers of firings representing other aspects of the situation, such as that person's movements, what she is saying, and so on. So if the experience seems much richer than just an orderly trip up a hierarchy of features, it is.

But the basic mechanism of going up a hierarchy of pattern recognizers in which each higher conceptual level represents a more abstract and more integrated concept remains valid. The flow of information downward is even greater, as each activated level of recognized pattern sends predictions to the next lower-level pattern recognizer of what it is likely to be encountering next. The apparent lushness of human experience is a result of the fact that all of the hundreds of millions of pattern recognizers in our neocortex are considering their inputs simultaneously.

In chapter 5 I'll discuss the flow of information from touch, vision, hearing, and other sensory organs into the neocortex. These early inputs are processed by cortical regions that are devoted to relevant types of sensory input (although there is enormous plasticity in the assignment of these regions, reflecting the basic uniformity of function in the neocortex). The conceptual hierarchy continues above the highest concepts in each sensory region of the neocortex. The cortical association areas integrate input from the different sensory inputs. When we hear something that perhaps sounds like our spouse's voice, and then see something that is perhaps indicative of her presence, we don't engage in an elaborate process of logical deduction; rather, we instantly perceive that our spouse is present from the combination of these sensory recognitions. We integrate all of the germane sensory and perceptual cues—perhaps even the smell of her perfume or his cologne—as one multilevel perception.

At a conceptual level above the cortical sensory association areas, we are capable of dealing with—perceiving, remembering, and thinking about—even more abstract concepts. At the highest level we recognize patterns such as *that's funny*, or *she's pretty*, or *that's ironic*, and so on. Our memories include these abstract recognition patterns as well. For example, we might recall that we were taking a walk with someone and that she said something funny, and we laughed, though we may not remember the actual joke itself. The memory sequence for that recollection has simply recorded the perception of humor but not the precise content of what was funny.

In the previous chapter I noted that we can often recognize a pattern even though we don't recognize it well enough to be able to describe it. For example, I believe I could pick out a picture of the woman with the baby carriage whom I saw earlier today from among a group of pictures of other women, despite the fact that I am unable to actually visualize her and cannot describe much specific about her. In this case my memory of her is a list of certain high-level features. These features do not have language or image labels attached to them, and they are not pixel images, so while I am able to think about her, I am unable to describe her. However, if I am presented with a picture of her, I can process the image, which results in the recognition of the same high-level features that were recognized the first time I saw her. I would be able to thereby determine that the features match and thus confidently pick out her picture.

Even though I saw this woman only once on my walk, there are probably already multiple copies of her pattern in my neocortex. However, if I don't think about her for a given period of time, then these pattern recognizers will become reassigned to other patterns. That is why memories grow dimmer with time: The amount of redundancy becomes reduced until certain memories become extinct. However, now that I have memorialized this particular woman by writing about her here, I probably won't forget her so easily.

Autoassociation and Invariance

In the previous chapter I discussed how we can recognize a pattern even if the entire pattern is not present, and also if it is distorted. The first capability is called autoassociation: the ability to associate a pattern with a part of itself. The structure of each pattern recognizer inherently supports this capability.

As each input from a lower-level pattern recognizer flows up to a higher-level one, the connection can have a "weight," indicating how important that particular element in the pattern is. Thus the more significant elements of a pattern are more heavily weighted in considering whether that pattern should trigger as "recognized." Lincoln's beard, Elvis's sideburns, and Einstein's famous tongue gesture are likely to have high weights in the patterns we've learned about the appearance of these iconic figures. The pattern recognizer computes a probability that takes the importance parameters into account. Thus the overall probability is lower if one or more of the elements is missing, though the threshold of recognition may nonetheless be met. As I pointed out, the computation of the overall probability (that the pattern is present) is more complicated than a simple weighted sum in that the size parameters also need to be considered.

If the pattern recognizer has received a signal from a higher-level recognizer that its pattern is "expected," then the threshold is effectively lowered (that is, made easier to achieve). Alternatively, such a signal may simply add to the total of the weighted inputs, thereby compensating for a missing element. This happens at every level, so that a pattern such as a face that is several levels up from the bottom may be recognized even with multiple missing features.

The ability to recognize patterns even when aspects of them are transformed is called feature invariance, and is dealt with in four ways. First, there are global transformations that are accomplished before the neocortex receives sensory data. We will discuss the voyage of sensory data from the eyes, ears, and skin in the section "The Sensory Pathway" on page 94.

The second method takes advantage of the redundancy in our cortical pattern memory. Especially for important items, we have learned many different perspectives and vantage points for each pattern. Thus many variations are separately stored and processed.

The third and most powerful method is the ability to combine two lists. One list can have a set of transformations that we have learned may apply to a certain category of pattern; the cortex will apply this same list of possible changes to another pattern. That is how we understand such language phenomena as metaphors and similes.

For example, we have learned that certain phonemes (the basic sounds of language) may be missing in spoken speech (for example, "goin'"). If we then learn a new spoken word (for example, "driving"), we will be able to recognize that word if one of its phonemes is missing even if we have never experienced that word in that form before, because we have become familiar with the general phenomenon of certain phonemes being omitted. As another example, we may learn that a particular artist likes to emphasize (by making larger) certain elements of a face, such as the nose. We can then identify a face with which we are familiar to which that modification has been applied even if we have never seen that modification on that face. Certain artistic modifications emphasize the very features that are recognized by our pattern recognition–based neocortex. As mentioned, that is precisely the basis of caricature.

The fourth method derives from the size parameters that allow a single module to encode multiple instances of a pattern. For example, we have heard the word "steep" many times. A particular pattern recognition module that is recognizing this spoken word can encode these multiple examples by indicating that the duration of [E] has a high expected variability. If all the modules for words including [E] share a similar phenomenon, that variability could be encoded in the models for [E] itself. However, different words incorporating [E] (or many other phonemes) may have different amounts of expected variability. For example, the word "peak" is likely not to have the [E] phoneme as drawn out as in the word "steep."

Learning

Are we not ourselves creating our successors in the supremacy of the earth? Daily adding to the beauty and delicacy of their organization,

daily giving them greater skill and supplying more and more of that self-regulating self-acting power which will be better than any intellect?
—Samuel Butler, 1871

The principal activities of brains are making changes in themselves.

—Marvin Minsky, *The Society of Mind*

So far we have examined how we recognize (sensory and perceptual) patterns and recall sequences of patterns (our memory of things, people, and events). However, we are not born with a neocortex filled with any of these patterns. Our neocortex is virgin territory when our brain is created. It has the capability of learning and therefore of creating connections between its pattern recognizers, but it gains those connections from experience.

This learning process begins even before we are born, occurring simultaneously with the biological process of actually growing a brain. A fetus already has a brain at one month, although it is essentially a reptile brain, as the fetus actually goes through a high-speed re-creation of biological evolution in the womb. The natal brain is distinctly a human brain with a human neocortex by the time it reaches the third trimester of pregnancy. At this time the fetus is having experiences, and the neocortex is learning. She can hear sounds, especially her mother's heartbeat, which is one likely reason that the rhythmic qualities of music are universal to human culture. Every human civilization ever discovered has had music as part of its culture, which is not the case with other art forms, such as pictorial art. It is also the case that the beat of music is comparable to our heart rate. Music beats certainly vary—otherwise music would not keep our interest—but heartbeats vary also. An overly regular heartbeat is actually a symptom of a diseased heart. The eyes of a fetus are partially open twenty-six weeks after conception, and are fully open most of the time by twenty-eight weeks after conception. There may not be much to see inside the womb, but there are patterns of light and dark that the neocortex begins to process.

So while a newborn baby has had a bit of experience in the womb, it is clearly limited. The neocortex may also learn from the old brain (a topic I discuss in chapter 5), but in general at birth the child has a lot to learn—everything from basic primitive sounds and shapes to metaphors and sarcasm.

Learning is critical to human intelligence. If we were to perfectly model and simulate the human neocortex (as the Blue Brain Project is attempting to do) and all of the other brain regions that it requires to function (such as the hippocampus and thalamus), it would not be able to do very much—in the same way that a newborn infant cannot do much (other than to be cute, which is definitely a key survival adaptation).

Learning and recognition take place simultaneously. We start learning immediately, and as soon as we've learned a pattern, we immediately start recognizing it. The neocortex is continually trying to make sense of the input presented to it. If a particular level is unable to fully process and recognize a pattern, it gets sent to the next higher level. If none of the levels succeeds in recognizing a pattern, it is deemed to be a new pattern. Classifying a pattern as new does not necessarily mean that every aspect of it is new. If we are looking at the paintings of a particular artist and see a cat's face with the nose of an elephant, we will be able to identify each of the distinctive features but will notice that this combined pattern is something novel, and are likely to remember it. Higher conceptual levels of the neocortex, which understand context—for example, the circumstance that this picture is an example of a particular artist's work and that we are attending an opening of a showing of new paintings by that artist—will note the unusual combination of patterns in the cat-elephant face but will also include these contextual details as additional memory patterns.

New memories such as the cat-elephant face are stored in an available pattern recognizer. The hippocampus plays a role in this process, and we'll discuss what is known about the actual biological mechanisms in the following chapter. For the purposes of our neocortex model, it is sufficient to say that patterns that are not otherwise recognized are stored as new patterns and are appropriately connected to the lower-level patterns that form them. The cat-elephant face, for example, will be stored in several different ways: The novel arrangement of facial parts will be stored as well as contextual memories that include the artist, the situation, and perhaps the fact that we laughed when we first saw it.

Memories that are successfully recognized may also result in the creation of a new pattern to achieve greater redundancy. If patterns are not perfectly recognized, they are likely to be stored as reflecting a different perspective of the item that was recognized.

What, then, is the overall method for determining what patterns get stored? In mathematical terms, the problem can be stated as follows: Using the available limits of pattern storage, how do we optimally represent the input patterns that have thus far been presented? While it makes sense to allow for a certain amount of redundancy, it would not be practical to fill up the entire available storage area (that is, the entire neocortex) with repeated patterns, as that would not allow for a sufficient diversity of patterns. A pattern such as the [E] phoneme in spoken words is something we have experienced countless times. It is a simple pattern of sound frequencies and it undoubtedly enjoys significant redundancy in our neocortex. We could fill up our entire neocortex with repeated patterns of the [E] phoneme. There is a limit, however, to useful redundancy, and a common pattern such as this clearly has reached it.

There is a mathematical solution to this optimization problem called linear programming, which solves for the best possible allocation of limited resources (in this case, a limited number of pattern recognizers) that would represent all of the cases on which the system has trained. Linear programming is designed for systems with one-dimensional inputs, which is another reason why it is optimal to represent the input to each pattern recognition module as a linear string of inputs. We can use this mathematical approach in a software system, and though an actual brain is further constrained by the physical connections it has available that it can adapt between pattern recognizers, the method is nonetheless similar.

An important implication of this optimal solution is that experiences that are routine are recognized but do not result in a permanent memory's being made. With regard to my walk, I experienced millions of patterns at every level, from basic visual edges and shadings to objects such as lampposts and mailboxes and people and animals and plants that I passed. Almost none of what I experienced was unique, and the patterns that I recognized had long since reached their optimal level of redundancy. The result is that I recall almost nothing from this walk. The few details that I do remember are likely to get overwritten with new patterns by the time I take another few dozen walks—except for the fact that I have now memorialized this particular walk by writing about it.

One important point that applies to both our biological neocortex and attempts to emulate it is that it is difficult to learn too many conceptual levels simultaneously. We can essentially learn one or at most two conceptual levels at a time. Once that learning is relatively stable, we can go on to learn the next level. We may continue to fine-tune the learning in the lower levels, but our learning focus is on the next level of abstraction. This is true at both the beginning of life, as newborns struggle with basic shapes, and later in life, as we struggle to learn new subject matter, one level of complexity at a time. We find the same phenomenon in machine emulations of the neocortex. However, if they are presented increasingly abstract material one level at a time, machines are capable of learning just as humans do (although not yet with as many conceptual levels).

The output of a pattern can feed back to a pattern at a lower level or even to the pattern itself, giving the human brain its powerful recursive ability. An element of a pattern can be a decision point based on another pattern. This is especially useful for lists that compose actions—for example, getting another tube of toothpaste if the current one is empty. These conditionals exist at every level. As anyone who has attempted to program a procedure on a computer knows, conditionals are vital to describing a course of action.

The Language of Thought

The dream acts as a safety-valve for the over-burdened brain.

—Sigmund Freud,

The Interpretation of Dreams, 1911

Brain: an apparatus with which we think we think.

—Ambrose Bierce, *The Devil's Dictionary*

To summarize what we've learned so far about the way the neocortex works, please refer to the diagram of the neocortical pattern recognition module on page 42.

a) Dendrites enter the module that represents the pattern. Even though patterns may seem to have two- or three-dimensional qualities, they are represented by a one-dimensional sequence of signals. The pattern must be present in this (sequential) order for the pattern recognizer to be able to recognize it. Each of the dendrites is connected ultimately to one or more axons of pattern recognizers at a lower conceptual level that have recognized a lower-level pattern that constitutes part of this pattern. For each of these input patterns, there may be many lower-level pattern recognizers that can generate the signal that the lower-level pattern has been recognized. The necessary threshold to recognize the pattern may be achieved even if not all of the inputs have signaled. The module computes the probability that the pattern it is responsible for is present. This computation considers the "importance" and "size" parameters (see [f] below).

Note that some of the dendrites transmit signals into the module and some out of the module. If all of the input dendrites to this pattern recognizer are signaling that their lower-level patterns have been recognized except for one or two, then this pattern recognizer will send a signal down to the pattern recognizer(s) recognizing the lower-level patterns that have not yet been recognized, indicating that there is a high likelihood that

that pattern will soon be recognized and that lower-level recognizer(s) should be on the lookout for it.

b) When this pattern recognizer recognizes its pattern (based on all or most of the input dendrite signals being activated), the axon (output) of this pattern recognizer will activate. In turn, this axon can connect to an entire network of dendrites connecting to many higher-level pattern recognizers that this pattern is input to. This signal will transmit magnitude information so that the pattern recognizers at the next higher conceptual level can consider it.

c) If a higher-level pattern recognizer is receiving a positive signal from all or most of its constituent patterns except for the one represented by this pattern recognizer, then that higher-level recognizer might send a signal down to this recognizer indicating that its pattern is expected. Such a signal would cause this pattern recognizer to lower its threshold, meaning that it would be more likely to send a signal on its axon (indicating that its pattern is considered to have been recognized) even if some of its inputs are missing or unclear.

d) Inhibitory signals from below would make it less likely that this pattern recognizer will recognize its pattern. This can result from recognition of lower-level patterns that are inconsistent with the pattern associated with this pattern recognizer (for example, recognition of a mustache by a lower-level recognizer would make it less likely that this image is "my wife").

e) Inhibitory signals from above would also make it less likely that this pattern recognizer will recognize its pattern. This can result from a higher-level context that is inconsistent with the pattern associated with this recognizer.

f) For each input, there are stored parameters for importance, expected size, and expected variability of size. The module computes an overall probability that the pattern is present based on all of these parameters and the current signals indicating which of the inputs are present and their magnitudes. A mathematically optimal way to accomplish this is with a technique called hidden Markov models. When such models are organized in a hierarchy (as they are in the neocortex or in attempts to simulate a neocortex), we call them hierarchical hidden Markov models.

Patterns triggered in the neocortex trigger other patterns. Partially complete patterns send signals down the conceptual hierarchy; completed patterns send signals up the conceptual hierarchy. These neocortical patterns are the language of thought. Just like language, they are hierarchical, but they are not language per se. Our thoughts are not conceived primarily in the elements of language, although since language also exists as hierarchies of patterns in our neocortex, we can have language-based thoughts. But for the most part, thoughts are represented in these neocortical patterns.

As I discussed above, if we were able to detect the pattern activations in someone's neocortex, we would still have little idea what those pattern activations meant without also having access to the entire hierarchy of patterns above and below each activated pattern. That would pretty much require access to that person's entire neocortex. It is hard enough for us to understand the content of our own thoughts, but understanding another person's requires mastering a neocortex different from our own. Of course we don't yet have access to someone else's neocortex; we need instead to rely on her attempts to express her thoughts into language (as well as other means such as gestures). People's incomplete ability to accomplish these communication tasks adds another layer of complexity—it is no wonder that we misunderstand one another as much as we do.

We have two modes of thinking. One is nondirected thinking, in which thoughts trigger one another in a nonlogical way. When we experience a sudden recollection of a memory from years or decades ago while doing something else, such as raking the leaves or walking down the street, the experience is recalled—as all memories are—as a sequence of patterns. We do not immediately visualize the scene unless we can call upon a lot of other memories that enable us to synthesize a more robust recollection. If we do visualize the scene in that way, we are essentially creating it in our mind from hints at the time of recollection; the memory itself is not stored in the form of images or visualizations. As I mentioned earlier, the triggers that led this thought to pop into our mind may or may not be evident. The sequence of relevant thoughts may have been immediately forgotten. Even if we do remember it, it will be a nonlinear and circuitous sequence of associations.

The second mode of thinking is directed thinking, which we use when we attempt to solve a problem or formulate an organized response. For example, we might be rehearsing in our mind something we plan to say to someone, or we might be formulating a passage we want to write (in a book on the mind, perhaps). As we think about tasks such as these, we have already broken down each one into a hierarchy of subtasks. Writing a book, for example, involves writing chapters; each chapter has sections; each section has paragraphs; each paragraph contains sentences that express ideas; each idea has its configuration of elements; each element and each relationship between elements is an idea that needs to be articulated; and so on. At the same time, our neocortical structures have learned certain rules that should be followed. If the task is writing, then we should try to avoid unnecessary repetition; we should try to make sure that the reader can follow what is being written; we should try to follow rules about grammar and style; and so on. The writer needs therefore to build a model of the reader in his mind, and that construct is hierarchical as well. In doing directed thinking, we are stepping through lists in our neocortex, each of which expands into extensive hierarchies of sublists, each with its own considerations. Keep in mind that elements in a list in a neocortical pattern can include conditionals, so our subsequent thoughts and actions will depend on assessments made as we go through the process.

Moreover, each such directed thought will trigger hierarchies of undirected thoughts. A continual storm of ruminations attends both our sensory experiences and our attempts at directed thinking. Our actual mental experience is complex and messy, made up of these lightning storms of triggered patterns, which change about a hundred times a second.

The Language of Dreams

Dreams are examples of undirected thoughts. They make a certain amount of sense because the phenomenon of one thought's triggering another is based on the actual linkages of patterns in our neocortex. To the extent that a dream does not make sense, we attempt to fix it through our ability to confabulate. As I will describe in chapter 9, split-brain patients (whose corpus callosum, which connects the two hemispheres of the brain, is severed or damaged) will confabulate (make up) explanations with their left brain—which controls the speech center—to explain what the right brain just did with input that the left brain did not have access to. We confabulate all the time in explaining the outcome of events. If you want a good example of this, just tune in to the daily commentary on the movement of financial markets. No matter how the markets perform, it's always possible to come up with a good explanation for why it happened, and such after-the-fact commentary is plentiful. Of course, if these commentators really understood the markets, they wouldn't have to waste their time doing commentary.

The act of confabulating is of course also done in the neocortex, which is good at coming up with stories and explanations that meet certain constraints. We do that whenever we retell a story. We will fill in details that may not be available or that we may have forgotten so that the story makes more sense. That is why stories change over time as they are told over and over again by new storytellers with perhaps different agendas. As spoken language led to written language, however, we had a technology that could record a definitive version of a story and prevent this sort of drift.

The actual content of a dream, to the extent that we remember it, is again a sequence of patterns. These patterns represent constraints in a story; we then confabulate a story that fits these constraints. The version of the dream that we retell (even if only to ourselves silently) is this confabulation. As we recount a dream we trigger cascades of patterns that fill in the actual dream as we originally experienced it.

There is one key difference between dream thoughts and our thinking while awake. One of the lessons we learn in life is that certain actions, even thoughts, are not permissible in the real world. For example, we learn that we cannot immediately fulfill our desires. There are rules against grabbing the money in the cash register at a store, and constraints on interacting with a person to whom we may be physically attracted. We also learn that certain thoughts are not permissible because they are culturally forbidden. As we learn professional skills, we learn the ways of thinking that are recognized and rewarded in our professions, and thereby avoid patterns of thought that might betray the methods and norms of that profession. Many of these taboos are worthwhile, as they enforce social order and consolidate progress. However, they can also prevent progress by enforcing an unproductive orthodoxy. Such orthodoxy is precisely what Einstein left behind when he tried to ride a light beam with his thought experiments.

Cultural rules are enforced in the neocortex with help from the old brain, especially the amygdala. Every thought we have triggers other thoughts, and some of them will relate to associated dangers. We learn, for example, that breaking a cultural norm even in our private thoughts can lead to ostracism, which the neocortex realizes threatens our well-being. If we entertain such thoughts, the amygdala is triggered, and that generates fear, which generally leads to terminating that thought.

In dreams, however, these taboos are relaxed, and we will often dream about matters that are culturally, sexually, or professionally forbidden. It is as if our brain realizes that we are not an actual actor in the world while dreaming. Freud wrote about this phenomenon but also noted that we will disguise such dangerous thoughts, at least when we attempt to recall them, so that the awake brain continues to be protected from them.

Relaxing professional taboos turns out to be useful for creative problem solving. I use a mental technique each night in which I think about a particular problem before I go to sleep. This triggers sequences of thoughts that will continue into my dreams. Once I am dreaming, I can think—*dream*—about solutions to the problem without the burden of the professional restraints I carry during the day. I can then access these dream thoughts in the morning while in an in-between state of dreaming and being awake, sometimes referred to as "lucid dreaming."[5]

Freud also famously wrote about the ability to gain insight into a person's psychology by interpreting dreams. There is of course a vast literature on all aspects of this theory, but the fundamental notion of gaining insight into ourselves through examination of our dreams makes sense. Our dreams are created by our neocortex, and thus their substance can be revealing of the content and connections found there. The relaxation of the constraints on our thinking that exist while we are awake is also useful in revealing neocortical content that we otherwise would be unable to access directly. It is also reasonable to conclude that the patterns that end up in our dreams represent important matters to us and thereby clues in understanding our unresolved desires and fears.

The Roots of the Model

As I mentioned above, I led a team in the 1980s and 1990s that developed the technique of hierarchical hidden Markov models to recognize human speech and understand natural-language statements. This work was the predecessor to today's widespread commercial systems that recognize and understand what we are trying to tell them (car navigation systems that you can talk to, Siri on the iPhone, Google Voice Search, and many others). The technique we developed had substantially all of the attributes that I describe in the PRTM. It included a hierarchy of patterns with each higher level being conceptually more abstract than the one below it. For example, in speech recognition the levels included basic patterns of sound frequency at the lowest level, then phonemes, then words and phrases (which were often recognized as if they were words). Some of our speech recognition systems could understand the meaning of natural-language commands, so yet higher levels included such structures as noun and verb phrases. Each pattern recognition module could recognize a linear sequence of patterns from a lower conceptual level. Each input had parameters for importance, size, and variability of size. There were "downward" signals indicating that a lower-level pattern was expected. I discuss this research in more detail in chapter 7.

In 2003 and 2004, PalmPilot inventor Jeff Hawkins and Dileep George developed a hierarchical cortical model called hierarchical temporal memory. With science writer Sandra Blakeslee, Hawkins described this model eloquently in their book *On Intelligence*. Hawkins provides a strong case for the uniformity of the cortical algorithm and its hierarchical and list-based organization. There are some important differences between the model presented in *On Intelligence* and what I present in this book. As the name implies, Hawkins is emphasizing the temporal (time-based) nature of the constituent lists. In other words, the direction of the lists is always forward in time. His explanation for how the features in a two-dimensional pattern such as the printed letter "A" have a direction in time is predicated on eye movement. He explains that we visualize images using saccades, which are very rapid movements of the eye of which we are unaware. The information reaching the neocortex is therefore not a two-dimensional set of features but rather a time-ordered list. While it is true that our eyes do make very rapid movements, the sequence in which they view the features of a pattern such as the letter "A" does not always occur in a consistent temporal order. (For example, eye saccades will not always register the top vertex in "A" before its bottom concavity.) Moreover, we can recognize a visual pattern that is presented for only a few tens of milliseconds, which is too short a period of time for eye saccades to scan it. It is true that the pattern recognizers in the neocortex store a pattern as a list and that the list is indeed ordered, but the order does not necessarily represent time. That is often indeed the case, but it may also represent a spatial or higher-level conceptual ordering as I discussed above.

The most important difference is the set of parameters that I have included for each input into the pattern recognition module, especially the size and size variability parameters. In the 1980s we actually tried to recognize human speech without this type of information. This was motivated by linguists' telling us that the duration information was not especially important. This perspective is illustrated by dictionaries that write out the pronunciation of each word as a string of phonemes, for example the word "steep" as [s] [t] [E] [p], with no indication of how long each phoneme is expected to last. The implication is that if we create programs to recognize phonemes and then encounter this particular sequence of four phonemes (in a spoken utterance), we should be able to recognize that spoken word. The system we built using this approach worked to some extent but not well enough to deal with such attributes as a large vocabulary, multiple speakers, and words spoken continuously without pauses. When we used the technique of hierarchical hidden Markov models in order to incorporate the distribution of magnitudes of each input, performance soared.

CHAPTER 4

THE BIOLOGICAL NEOCORTEX

Because important things go in a case, you've got a skull for your brain, a plastic sleeve for your comb, and a wallet for your money.

—George Costanza, in "The Reverse Peephole"
episode of *Seinfeld*

Now, for the first time, we are observing the brain at work in a global manner with such clarity that we should be able to discover the overall programs behind its magnificent powers.

—J. G. Taylor, B. Horwitz, and K. J. Friston

The mind, in short, works on the data it receives very much as a sculptor works on his block of stone. In a sense the statue stood there from eternity. But there were a thousand different ones beside it, and the sculptor alone is to thank for having extricated this one from the rest. Just so the world of each of us, howsoever different our several views of it may be, all lay embedded in the primordial chaos of sensations, which gave the mere *matter* to the thought of all of us indifferently. We may, if we like, by our reasonings unwind things back to that black and jointless continuity of space and moving clouds of swarming atoms which science calls the only real world. But all the while the world *we* feel and live in will be that which our ancestors and we, by slowly cumulative strokes of choice, have extricated out of this, like sculptors, by simply rejecting certain portions of the given stuff. Other sculptors, other statues from the same stone! Other minds, other worlds from the same monotonous and inexpressive chaos! My world is but one in a million alike embedded, alike real to those who may abstract them. How different must be the worlds in the consciousness of ant, cuttle-fish, or crab!

—William James

Is intelligence the goal, or even *a* goal, of biological evolution? Steven Pinker writes, "We are chauvinistic about our brains, thinking them to be the goal of evolution,"[1] and goes on to argue that "that makes no sense. . . . Natural selection does nothing even close to striving for intelligence. The process is driven by differences in the survival and reproduction rates of replicating organisms in a particular environment. Over time, the organisms acquire designs that adapt them for survival and reproduction in that environment, period; nothing pulls them in any direction other than success there and then." Pinker concludes that "life is a densely branching bush, not a scale or a ladder, and living organisms are at the tips of branches, not on lower rungs."

With regard to the human brain, he questions whether the "benefits outweigh the costs." Among the costs, he cites that "the brain [is] bulky. The female pelvis barely accommodates a baby's outsized head. That design compromise kills many women during childbirth and requires a pivoting gait that makes women biomechanically less efficient walkers than men. Also a heavy head bobbing around on a neck makes us more vulnerable to fatal injuries in accidents such as falls." He goes on to list additional shortcomings, including the brain's energy consumption, its slow reaction time, and the lengthy process of learning.

While each of these statements is accurate on its face (although many

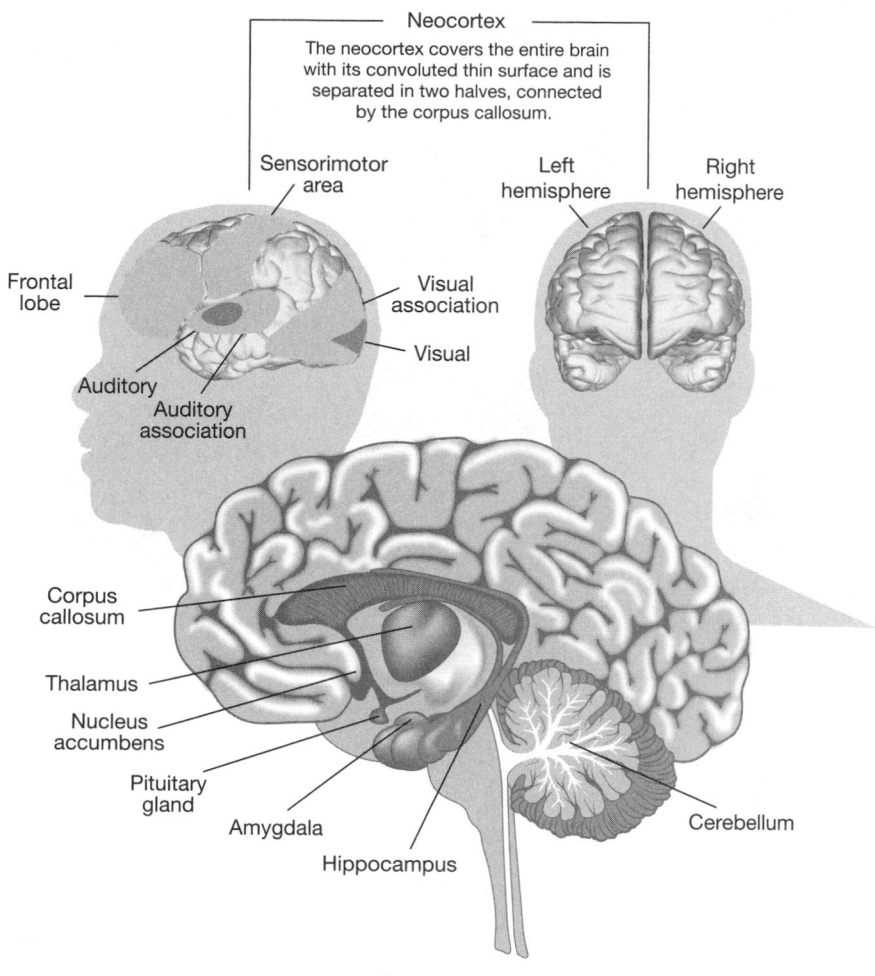

Physical layout of key regions of the brain.

of my female friends are better walkers than I am), Pinker is missing the overall point here. It is true that biologically, evolution has no specific direction. It is a search method that indeed thoroughly fills out the "densely branching bush" of nature. It is likewise true that evolutionary changes do not *necessarily* move in the direction of greater intelligence—they move in *all* directions. There are many examples of successful creatures

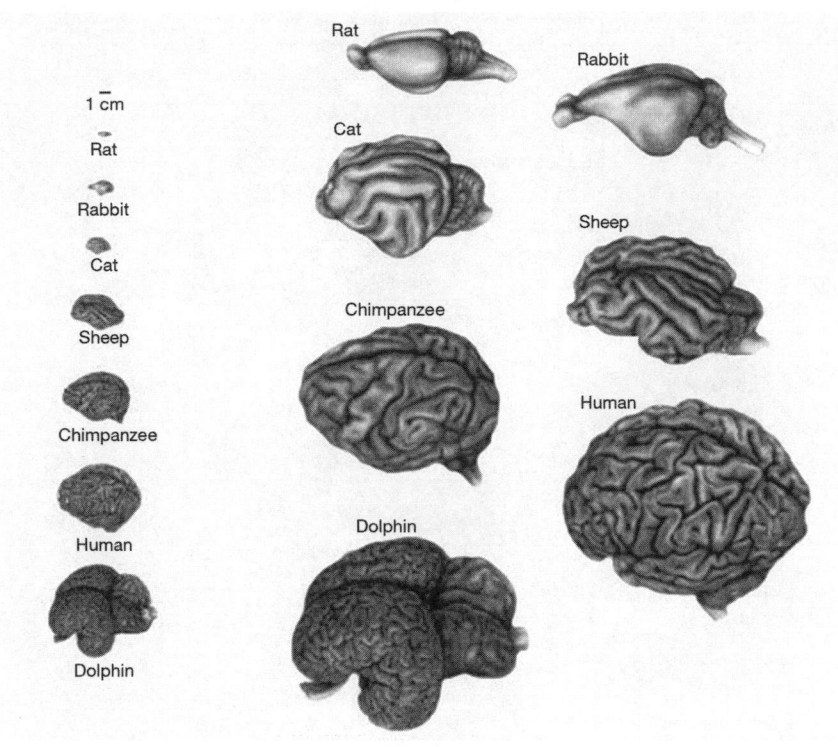

The neocortex in different mammals.

that have remained relatively unchanged for millions of years. (Alligators, for instance, date back 200 million years, and many microorganisms go back much further than that.) But in the course of thoroughly filling out myriad evolutionary branches, one of the directions it *does* move in is toward greater intelligence. That is the relevant point for the purposes of this discussion.

Suppose we have a blue gas in a jar. When we remove the lid, there is no message that goes out to all of the molecules of the gas saying, "Hey, guys, the lid is off the jar; let's head up toward the opening and out to freedom." The molecules just keep doing what they always do, which is to move every which way with no seeming direction. But in the course of doing so, some of them near the top will indeed move out of the jar, and over time most of them will follow suit. Once biological evolution stumbled on a neural mechanism capable of hierarchical learning, it found it to be immensely useful for evolution's one objective, which is survival. The benefit of having a neocortex became acute when quickly changing circumstances favored rapid learning. Species of all kinds—plants and animals— can learn to adapt to changing circumstances over time, but without a neocortex they must use the process of genetic evolution. It can take a great many generations—thousands of years—for a species without a neocortex to learn significant new behaviors (or in the case of plants, other adaptation strategies). The salient survival advantage of the neocortex was that it could learn in a matter of days. If a species encounters dramatically changed circumstances and one member of that species invents or discovers or just stumbles upon (these three methods all being variations of innovation) a way to adapt to that change, other individuals will notice, learn, and copy that method, and it will quickly spread virally to the entire population. The cataclysmic Cretaceous-Paleogene extinction event about 65 million years ago led to the rapid demise of many non-neocortex-bearing species that could not adapt quickly enough to a suddenly altered environment. This marked the turning point for neocortex-capable mammals to take over their ecological niche. In this way, biological evolution found that the hierarchical learning of the neocortex was so valuable that this region of the brain continued to grow in size until it virtually took over the brain of *Homo sapiens*.

Discoveries in neuroscience have established convincingly the key role played by the hierarchical capabilities of the neocortex as well as offered evidence for the pattern recognition theory of mind (PRTM). This evidence is distributed among many observations and analyses, a portion of which I will review here. Canadian psychologist Donald O. Hebb (1904– 1985) made an initial attempt to explain the neurological basis of learning. In 1949 he described a mechanism in which neurons change physiologically based on their experience, thereby providing a basis for learning and brain plasticity: "Let us assume that the persistence or repetition of a reverberatory activity (or 'trace') tends to induce lasting cellular changes that add to its stability. . . . When an axon of cell *A* is near enough to excite a cell *B* and repeatedly or persistently takes part in firing it, some growth process or metabolic change takes place in one or both cells such that *A*'s efficiency, as one of the cells firing *B*, is increased."[2] This theory has been stated as "cells that fire together wire together" and has become known as Hebbian learning. Aspects of Hebb's theory have been confirmed, in that it is clear that brain assemblies can create new connections and strengthen them, based on their own activity. We can actually see neurons developing such connections in brain scans. Artificial "neural nets" are based on Hebb's model of neuronal learning.

The central assumption in Hebb's theory is that the basic unit of learning in the neocortex is the neuron. The pattern recognition theory of mind that I articulate in this book is based on a different fundamental unit: not the neuron itself, but rather an assembly of neurons, which I estimate to number around a hundred. The wiring and synaptic strengths *within* each unit are relatively stable and determined genetically—that is, the organization within each pattern recognition module is determined by genetic design. Learning takes place in the creation of connections *between* these units, not within them, and probably in the synaptic strengths of those interunit connections.

Recent support for the basic module of learning's being a module of dozens of neurons comes from Swiss neuroscientist Henry Markram (born in 1962), whose ambitious Blue Brain Project to simulate the entire human brain I describe in chapter 7. In a 2011 paper he describes how while scanning and analyzing actual mammalian neocortex neurons, he was "search[ing] for evidence of Hebbian assemblies at the most elementary level of the cortex." What he found instead, he writes, were "elusive assemblies [whose] connectivity and synaptic weights are highly predictable and constrained." He concludes that "these findings imply that experience cannot easily mold the synaptic connections of these assemblies" and speculates that "they serve as innate, Lego-like building blocks of knowledge for perception and that the acquisition of memories involves the combination of these building blocks into complex constructs." He continues:

> Functional neuronal assemblies have been reported for decades, but direct evidence of clusters of synaptically connected neurons . . . has been missing. . . . Since these assemblies will all be similar in topology and synaptic weights, not molded by any specific experience, we consider these to be innate assemblies. . . . Experience plays only a minor role in determining synaptic connections and weights within these assemblies. . . . Our study found evidence [of] innate Lego-like assemblies of a few dozen neurons. . . . Connections between assemblies may combine them into super-assemblies within a neocortical layer, then in higher-order assemblies in a cortical column, even higher-order assemblies in a brain region, and finally in the highest possible order assembly represented by the whole brain. . . . Acquiring memories is very similar to building with Lego. Each assembly is equivalent to a Lego block holding some piece of elementary innate knowledge about how to process, perceive and respond to the world. . . . When different blocks come together, they therefore form a unique combination of these innate percepts that represents an individual's specific knowledge and experience.[3]

The "Lego blocks" that Markram proposes are fully consistent with the pattern recognition modules that I have described. In an e-mail communication, Markram described these "Lego blocks" as "shared content and innate knowledge."[4] I would articulate that the purpose of these modules is to recognize patterns, to remember them, and to predict them based on partial patterns. Note that Markram's estimate of each module's containing "several dozen neurons" is based only on layer V of the neocortex. Layer V is indeed neuron rich, but based on the usual ratio of neuron counts in the six layers, this would translate to an order of magnitude of about 100 neurons per module, which is consistent with my estimates.

The consistent wiring and apparent modularity of the neocortex has been noted for many years, but this study is the first to demonstrate the stability of these modules as the brain undergoes its dynamic processes.

Another recent study, this one from Massachusetts General Hospital, funded by the National Institutes of Health and the National Science Foundation and published in a March 2012 issue of the journal *Science,* also shows a regular structure of connections across the neocortex.[5] The article describes the wiring of the neocortex as following a grid pattern, like orderly city streets: "Basically, the overall structure of the brain ends up resembling Manhattan, where you have a 2-D plan of streets and a third axis, an elevator going in the third dimension," wrote Van J. Wedeen, a Harvard neuroscientist and physicist and the head of the study.

In a *Science* magazine podcast, Wedeen described the significance of the research: "This was an investigation of the three-dimensional structure of the pathways of the brain. When scientists have thought about the pathways of the brain for the last hundred years or so, the typical image or model that comes to mind is that these pathways might resemble a bowl of spaghetti—separate pathways that have little particular spatial pattern in relation to one another. Using magnetic resonance imaging, we were able to investigate this question experimentally. And what we found was that rather than being haphazardly arranged or independent pathways, we find that all of the pathways of the brain taken together fit together in a single exceedingly simple structure. They basically look like a cube. They basically run in three perpendicular directions, and in each one of those three directions the pathways are highly parallel to each other and arranged in arrays. So, instead of independent spaghettis, we see that the connectivity of the brain is, in a sense, a single coherent structure."

Whereas the Markram study shows a module of neurons that repeats itself across the neocortex, the Wedeen study demonstrates a remarkably orderly pattern of connections between modules. The brain starts out with a very large number of "connections-in-waiting" to which the pattern recognition modules can hook up. Thus if a given module wishes to connect to another, it does not need to grow an axon from one and a dendrite from the other to span the entire physical distance between them. It can simply harness one of these axonal connections-in-waiting and just hook up to the ends of the fiber. As Wedeen and his colleagues write, "The pathways of the brain follow a base-plan established by . . . early embryogenesis. Thus, the pathways of the mature brain present an image of these three primordial gradients, physically deformed by development." In other words, as we learn and have experiences, the pattern recognition modules of the neocortex are connecting to these preestablished connections that were created when we were embryos.

There is a type of electronic chip called a field programmable gate array (FPGA) that is based on a similar principle. The chip contains millions of modules that implement logic functions along with connections-in-waiting. At the time of use, these connections are either activated or deactivated (through electronic signals) to implement a particular capability.

In the neocortex, those long-distance connections that are not used are eventually pruned away, which is one reason why adapting a nearby region of the neocortex to compensate for one that has become damaged is not quite as effective as using the original region. According to the Wedeen study, the initial connections are extremely orderly and repetitive, just like the modules themselves, and their grid pattern is used to "guide connectivity" in the neocortex. This pattern was found in all of the primate and human brains studied and was evident across the neocortex, from regions that dealt with early sensory patterns up to higher-level emotions. Wedeen's *Science* journal article concluded that the "grid structure of cerebral pathways was pervasive, coherent, and continuous with the three principal axes of development." This again speaks to a common algorithm across all neocortical functions.

It has long been known that at least certain regions of the neocortex are hierarchical. The best-studied region is the visual cortex, which is separated into areas known as V1, V2, and MT (also known as V5). As we

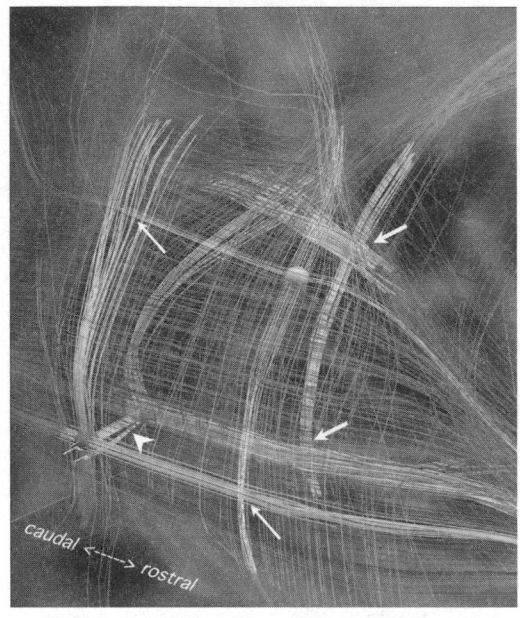

The highly regular grid structure of initial connections in the neocortex found in a National Institutes of Health study.

Another view of the regular grid structure of neocortical connections.

The grid structure found in the neocortex is remarkably similar to what is called crossbar switching, which is used in integrated circuits and circuit boards.

advance to higher areas in this region ("higher" in the sense of conceptual processing, not physically, as the neocortex is always just one pattern recognizer thick), the properties that can be recognized become more abstract. V1 recognizes very basic edges and primitive shapes. V2 can recognize contours, the disparity of images presented by each of the eyes, spatial orientation, and whether or not a portion of the image is part of an object or the background.[6] Higher-level regions of the neocortex recognize concepts such as the identity of objects and faces and their movement. It has also long been known that communication through this hierarchy is both upward and downward, and that signals can be both excitatory and inhibitory. MIT neuroscientist Tomaso Poggio (born in 1947) has extensively studied vision in the human brain, and his research for the last thirty-five years has been instrumental in establishing hierarchical learning and pattern recognition in the "early" (lowest conceptual) levels of the visual neocortex.[7]

Our understanding of the lower hierarchical levels of the visual neo-cortex is consistent with the PRTM I described in the previous chapter, and observation of the hierarchical nature of neocortical processing has recently extended far beyond these levels. University of Texas neurobiology professor Daniel J. Felleman and his colleagues traced the "hierarchical organization of the cerebral cortex . . . [in] 25 neocortical areas," which included both visual areas and higher-level areas that combine patterns from multiple senses. What they found as they went up the neocortical hierarchy was that the processing of patterns became more abstract, comprised larger spatial areas, and involved longer time periods. With every connection they found communication both up and down the hierarchy.[8]

Recent research allows us to substantially broaden these observations to regions well beyond the visual cortex and even to the association areas, which combine inputs from multiple senses. A study published in 2008 by Princeton psychology professor Uri Hasson and his colleagues demonstrates that the phenomena observed in the visual cortex occur across a wide variety of neocortical areas: "It is well established that neurons along the visual cortical pathways have increasingly larger spatial receptive fields. This is a basic organizing principle of the visual system. . . . Real-world events occur not only over extended regions of space, but also over extended periods of time. We therefore hypothesized that a hierarchy analogous to that found for spatial receptive field sizes should also exist for the temporal response characteristics of different brain regions." This is exactly what they found, which enabled them to conclude that "similar to the known cortical hierarchy of spatial receptive fields, there is a hierarchy of progressively longer temporal receptive windows in the human brain."[9]

The most powerful argument for the universality of processing in the neocortex is the pervasive evidence of plasticity (not just learning but interchangeability): In other words, one region is able to do the work of other regions, implying a common algorithm across the entire neocortex. A great deal of neuroscience research has been focused on identifying which regions of the neocortex are responsible for which types of patterns.

The classical technique for determining this has been to take advantage of brain damage from injury or stroke and to correlate lost functionality with specific damaged regions. So, for example, when we notice that someone with newly acquired damage to the fusiform gyrus region suddenly has difficulty recognizing faces but is still able to identify people from their voices and language patterns, we can hypothesize that this region has something to do with face recognition. The underlying assumption has been that each of these regions is designed to recognize and process a particular type of pattern. Particular physical regions have become associated with particular types of patterns, because under normal circumstances that is how the information happens to flow. But when that normal flow of information is disrupted for any reason, another region of the neocortex is able to step in and take over.

Plasticity has been widely noted by neurologists, who observed that patients with brain damage from an injury or a stroke can relearn the same skills in another area of the neocortex. Perhaps the most dramatic example of plasticity is a 2011 study by American neuroscientist Marina Bedny and her colleagues on what happens to the visual cortex of congenitally blind people. The common wisdom has been that the early layers of the visual cortex, such as V1 and V2, inherently deal with very low-level patterns (such as edges and curves), whereas the frontal cortex (that evolutionarily new region of the cortex that we have in our uniquely large foreheads) inherently deals with the far more complex and subtle patterns of language and other abstract concepts. But as Bedny and her colleagues found, "Humans are thought to have evolved brain regions in the left frontal and temporal cortex that are uniquely capable of language processing. However, congenitally blind individuals also activate the visual cortex in some verbal tasks. We provide evidence that this visual cortex activity in fact reflects language processing. We find that in congenitally blind individuals, the left visual cortex behaves similarly to classic language regions. . . . We conclude that brain regions that are thought to have evolved for vision can take on language processing as a result of early experience."[10]

Consider the implications of this study: It means that neocortical regions that are physically relatively far apart, and that have also been considered conceptually very different (primitive visual cues versus abstract language concepts), use essentially the same algorithm. The regions that process these disparate types of patterns can substitute for one another.

University of California at Berkeley neuroscientist Daniel E. Feldman wrote a comprehensive 2009 review of what he called "synaptic mechanisms for plasticity in the neocortex" and found evidence for this type of plasticity across the neocortex. He writes that "plasticity allows the brain to learn and remember patterns in the sensory world, to refine movements . . . and to recover function after injury." He adds that this plasticity is enabled by "structural changes including formation, removal, and morphological remodeling of cortical synapses and dendritic spines."[11]

Another startling example of neocortical plasticity (and therefore of the uniformity of the neocortical algorithm) was recently demonstrated by scientists at the University of California at Berkeley. They hooked up implanted microelectrode arrays to pick up brain signals specifically from a region of the motor cortex of mice that controls the movement of their whiskers. They set up their experiment so that the mice would get a reward if they controlled these neurons to fire in a certain mental pattern but not to actually move their whiskers. The pattern required to get the reward involved a mental task that their frontal neurons would normally not do. The mice were nonetheless able to perform this mental feat essentially by thinking with their motor neurons while mentally decoupling them from controlling motor movements.[12] The conclusion is that the motor cortex, the region of the neocortex responsible for coordinating muscle movement, also uses the standard neocortical algorithm.

There are several reasons, however, why a skill or an area of knowledge that has been relearned using a new area of the neocortex to replace one that has been damaged will not necessarily be as good as the original. First, because it took an entire lifetime to learn and perfect a given skill, relearning it in another area of the neocortex will not immediately generate the same results. More important, that new area of the neocortex has not just been sitting around waiting as a standby for an injured region. It too has been carrying out vital functions, and will therefore be hesitant to give up its neocortical patterns to compensate for the damaged region. It can start by releasing some of the redundant copies of its patterns, but doing so will subtly degrade its existing skills and does not free up as much cortical space as the skills being relearned had used originally.

There is a third reason why plasticity has its limits. Since in most people particular types of patterns will flow through specific regions (such as faces being processed by the fusiform gyrus), these regions have become optimized (by biological evolution) for those types of patterns. As I report in chapter 7, we found the same result in our digital neocortical developments. We could recognize speech with our character recognition systems and vice versa, but the speech systems were optimized for speech and similarly the character recognition systems were optimized for printed characters, so there would be some reduction in performance if we substituted one for the other. We actually used evolutionary (genetic) algorithms to accomplish this optimization, a simulation of what biology does naturally. Given that faces have been flowing through the fusiform gyrus for most people for hundreds of thousands of years (or more), biological evolution has had time to evolve a favorable ability to process such patterns in that region. It uses the same basic algorithm, but it is oriented toward faces. As Dutch neuroscientist Randal Koene wrote, "The [neo]cortex is very uniform, each column or minicolumn can in principle do what each other one can do."[13]

Substantial recent research supports the observation that the pattern recognition modules wire themselves based on the patterns to which they are exposed. For example, neuroscientist Yi Zuo and her colleagues watched as new "dendritic spines" formed connections between nerve cells as mice learned a new skill (reaching through a slot to grab a seed).[14] Researchers at the Salk Institute have discovered that this critical self-wiring of the neocortex modules is apparently controlled by only a handful of genes. These genes and this method of self-wiring are also uniform across the neocortex.[15]

Many other studies document these attributes of the neocortex, but let's summarize what we can observe from the neuroscience literature and from our own thought experiments. The basic unit of the neocortex is a module of neurons, which I estimate at around a hundred. These are woven together into each neocortical column so that each module is not visibly distinct. The pattern of connections and synaptic strengths within each module is relatively stable. It is the connections and synaptic strengths *between* modules that represent learning.

There are on the order of a quadrillion (10^{15}) connections in the neocortex, yet only about 25 million bytes of design information in the genome (after lossless compression),[16] so the connections themselves cannot possibly be predetermined genetically. It is possible that some of this learning is the product of the neocortex's interrogating the old brain, but that still would necessarily represent only a relatively small amount of information. The connections between modules are created on the whole from experience (nurture rather than nature).

The brain does not have sufficient flexibility so that each neocortical pattern recognition module can simply link to any other module (as we can easily program in our computers or on the Web)—an actual physical connection must be made, composed of an axon connecting to a dendrite. We each start out with a vast stockpile of possible neural connections. As the Wedeen study shows, these connections are organized in a very repetitive and orderly manner. Terminal connection to these axons-in-waiting takes place based on the patterns that each neocortical pattern recognizer has recognized. Unused connections are ultimately pruned away. These connections are built hierarchically, reflecting the natural hierarchical order of reality. That is the key strength of the neocortex.

The basic algorithm of the neocortical pattern recognition modules is equivalent across the neocortex from "low-level" modules, which deal with the most basic sensory patterns, to "high-level" modules, which recognize the most abstract concepts. The vast evidence of plasticity and the interchangeability of neocortical regions is testament to this important observation. There is some optimization of regions that deal with particular types of patterns, but this is a second-order effect—the fundamental algorithm is universal.

Signals go up and down the conceptual hierarchy. A signal going up means, "I've detected a pattern." A signal going down means, "I'm expecting your pattern to occur," and is essentially a prediction. Both upward and downward signals can be either excitatory or inhibitory.

Each pattern is itself in a particular order and is not readily reversed. Even if a pattern appears to have multidimensional aspects, it is represented by a one-dimensional sequence of lower-level patterns. A pattern is an ordered sequence of other patterns, so each recognizer is inherently recursive. There can be many levels of hierarchy.

There is a great deal of redundancy in the patterns we learn, especially the important ones. The recognition of patterns (such as common objects and faces) uses the same mechanism as our memories, which are just patterns we have learned. They are also stored as sequences of patterns—they are basically stories. That mechanism is also used for learning and carrying out physical movement in the world. The redundancy of patterns is what enables us to recognize objects, people, and ideas even when they have variations and occur in different contexts. The size and size variability parameters also allow the neocortex to encode variation in magnitude against different dimensions (duration in the case of sound). One way that these magnitude parameters could be encoded is simply through multiple patterns with different numbers of repeated inputs. So, for example, there could be patterns for the spoken word "steep" with different numbers of the long vowel [E] repeated, each with the importance parameter set to a moderate level indicating that the repetition of [E] is variable. This approach is not mathematically equivalent to having the explicit size parameters and does not work nearly as well in practice, but is one approach to encoding magnitude. The strongest evidence we have for these parameters is that they are needed in our AI systems to get accuracy levels that are near human levels.

The summary above constitutes the conclusions we can draw from the sampling of research results I have shared above as well as the sampling of thought experiments I discussed earlier. I maintain that the model I have presented is the only possible model that satisfies all of the constraints that the research and our thought experiments have established.

Finally, there is one more piece of corroborating evidence. The techniques that we have evolved over the past several decades in the field of artificial intelligence to recognize and intelligently process real-world phenomena (such as human speech and written language) and to understand natural-language documents turn out to be mathematically similar to the model I have presented above. They are also examples of the PRTM. The AI field was not explicitly trying to copy the brain, but it nonetheless arrived at essentially equivalent techniques.

CHAPTER 5

THE OLD BRAIN

I have an old brain but a terrific memory.

—Al Lewis

Here we stand in the middle of this new world with our primitive brain, attuned to the simple cave life, with terrific forces at our disposal, which we are clever enough to release, but whose consequences we cannot comprehend.

—Albert Szent-Györgyi

Our old brain—the one we had before we were mammals—has not disappeared. Indeed it still provides much of our motivation in seeking gratification and avoiding danger. These goals are modulated, however, by our neocortex, which dominates the human brain in both mass and activity.

Animals used to live and survive without a neocortex, and indeed all nonmammalian animals continue to do so today. We can view the human neocortex as the great sublimator—thus our primitive motivation to avoid a large predator may be transformed by the neocortex today into completing an assignment to impress our boss; the great hunt may become writing a book on, say, the mind; and pursuing reproduction may become gaining public recognition or decorating your apartment. (Well, this last motivation is not always so hidden.)

The neocortex is likewise good at helping us solve problems because it can accurately model the world, reflecting its true hierarchical nature. But it is the old brain that presents us with those problems. Of course, like any clever bureaucracy, the neocortex often deals with the problems it is assigned by redefining them. On that note, let's review the information processing in the old brain.

The Sensory Pathway

> Pictures, propagated by motion along the fibers of the optic nerves in the brain, are the cause of vision.
>
> **—Isaac Newton**

> Each of us lives within the universe—the prison—of his own brain. Projecting from it are millions of fragile sensory nerve fibers, in groups uniquely adapted to sample the energetic states of the world around us: heat, light, force, and chemical composition. That is all we ever know of it directly; all else is logical inference.
>
> **—Vernon Mountcastle[1]**

Although we experience the illusion of receiving high-resolution images from our eyes, what the optic nerve actually sends to the brain is just a series of outlines and clues about points of interest in our visual field. We then essentially hallucinate the world from cortical memories that interpret a series of movies with very low data rates that arrive in parallel channels. In a study published in *Nature,* Frank S. Werblin, professor of molecular and cell biology at the University of California at Berkeley, and doctoral student Boton Roska, MD, showed that the optic nerve carries ten to twelve

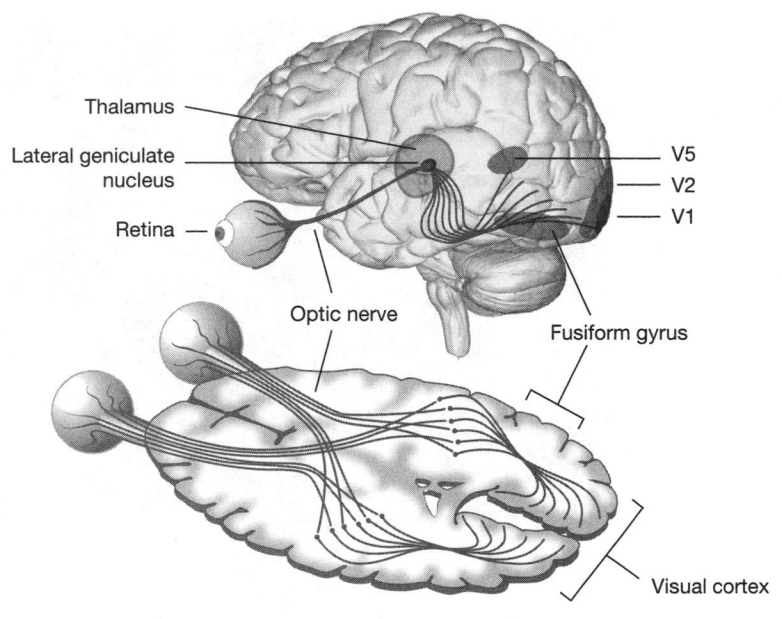

Thalamus

Lateral geniculate
nucleus

Retina

Optic nerve

V5

V2

V1

Fusiform gyrus

Visual cortex

The visual pathway in the brain.

output channels, each of which carries only a small amount of information about a given scene.[2] One group of what are called ganglion cells sends information only about edges (changes in contrast). Another group detects only large areas of uniform color, whereas a third group is sensitive only to the backgrounds behind figures of interest.

"Even though we think we see the world so fully, what we are receiving is really just hints, edges in space and time," says Werblin. "These 12 pictures of the world constitute all the information we will ever have about what's out there, and from these 12 pictures, which are so sparse, we reconstruct the richness of the visual world. I'm curious how nature selected these 12 simple movies and how it can be that they are sufficient to provide us with all the information we seem to need."

This data reduction is what in the AI field we call "sparse coding." We have found in creating artificial systems that throwing most of the input information away and retaining only the most salient details provides

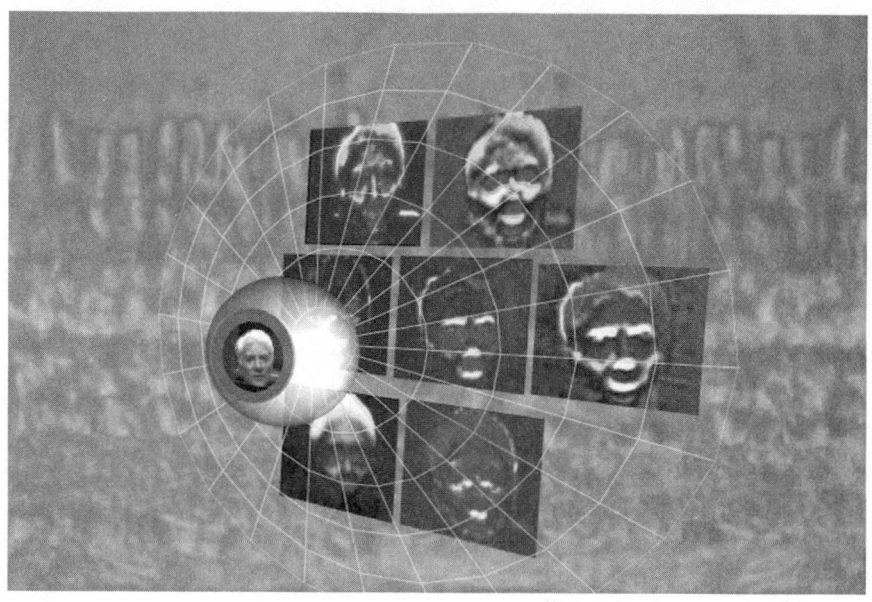

Seven of the twelve low-data-rate "movies" sent by the optic nerve to the brain.

superior results. Otherwise the limited ability to process information in a neocortex (biological or otherwise) gets overwhelmed.

The processing of auditory information from the human cochlea through the subcortical regions and then through the early stages of the neocortex has been meticulously modeled by Lloyd Watts and his research team at Audience, Inc.[3] They have developed research technology that extracts 600 different frequency bands (60 per octave) from sound. This comes much closer to the estimate of 3,000 bands extracted by the human cochlea (compared with commercial speech recognition, which uses only 16 to 32 bands). Using two microphones and its detailed (and high–spectral resolution) model of auditory processing, Audience has created a commercial technology (with somewhat lower spectral resolution than its research system) that effectively removes background noise from conversations. This is now being used in many popular cell phones and is an impressive

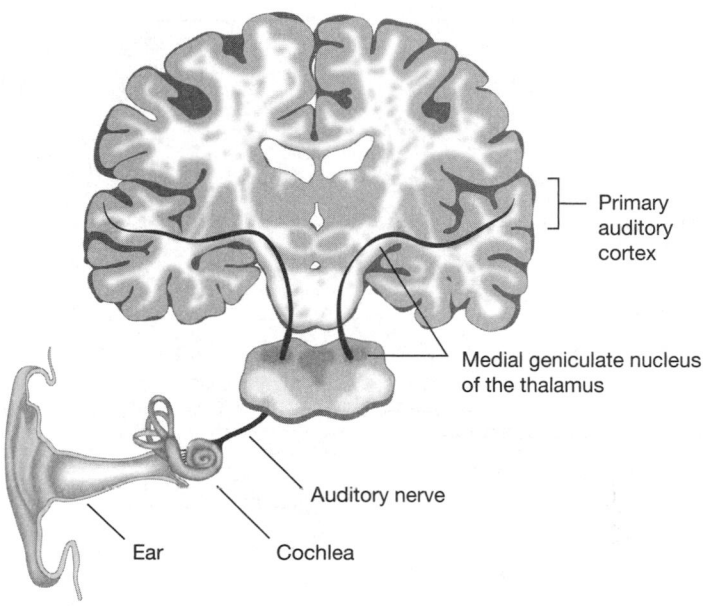

The auditory pathway in the brain.

example of a commercial product based on an understanding of how the human auditory perceptual system is able to focus on one sound source of interest.

Inputs from the body (estimated at hundreds of megabits per second), including that of nerves from the skin, muscles, organs, and other areas, stream into the upper spinal cord. These messages involve more than just communication about touch; in addition they carry information about temperature, acid levels (for example, lactic acid in muscles), the movement of food through the gastrointestinal tract, and many other signals. This data is processed through the brain stem and midbrain. Key cells called lamina 1 neurons create a map of the body, representing its current state, not unlike the displays used by flight controllers to track airplanes. From here the sensory data heads to a mysterious region called the thalamus, which brings us to our next topic.

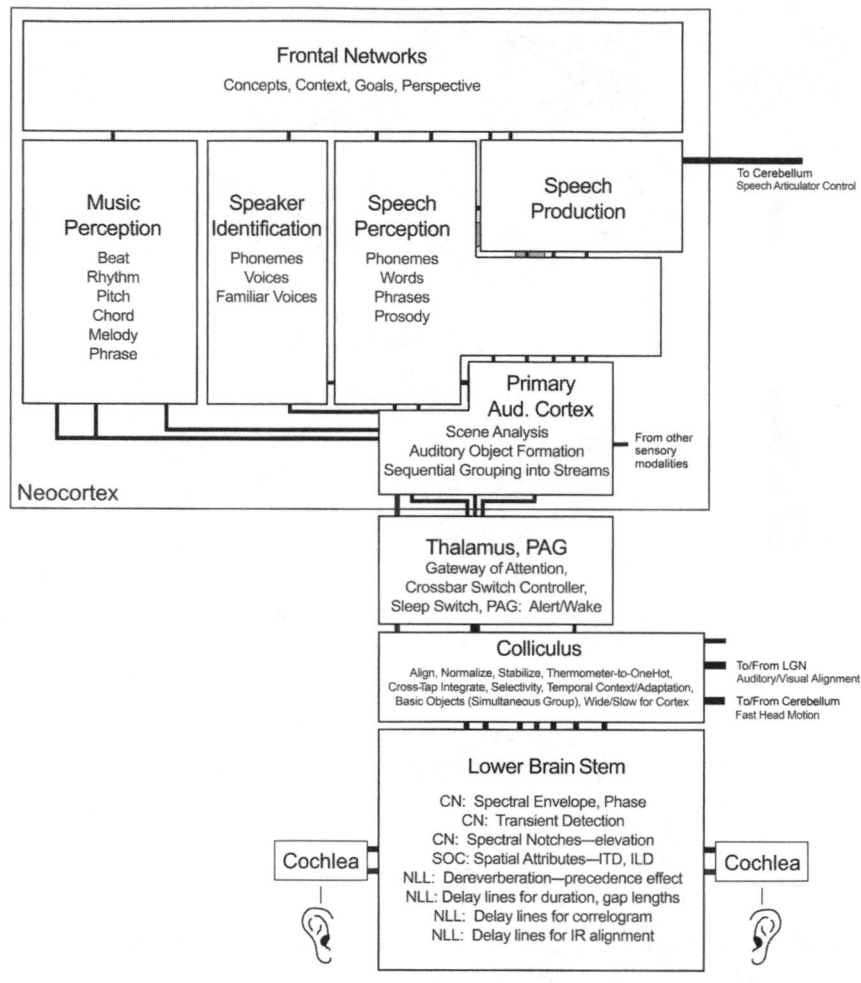

A simplified model of auditory processing in both the subcortical areas (areas prior to the neocortex) and the neocortex, created by Audience, Inc. Figure adapted from L. Watts, "Reverse-Engineering the Human Auditory Pathway," in J. Liu et al. (eds.), *WCCI 2012* (Berlin: Springer-Verlag, 2012), p. 49.

The Thalamus

Everyone knows what attention is. It is the taking possession by the mind, in clear and vivid form, of one out of what seem several simultaneously possible objects or trains of thought. Focalization, concentration, of consciousness, are of its essence. It implies withdrawal from some things in order to deal effectively with others.

—William James

From the midbrain, sensory information then flows through a nut-sized region called the posterior ventromedial nucleus (VMpo) of the thalamus, which computes complex reactions to bodily states such as "this tastes terrible," "what a stench," or "that light touch is stimulating." The increasingly processed information ends up at two regions of the neocortex called the insula. These structures, the size of small fingers, are located on the left and

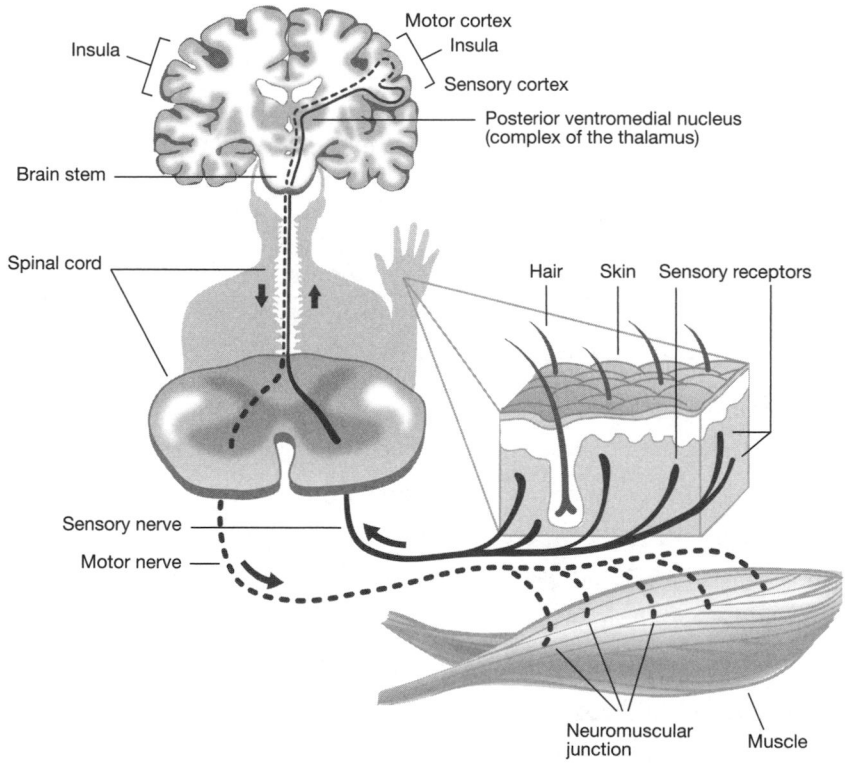

The sensory-touch pathway in the brain.

right sides of the neocortex. Dr. Arthur Craig of the Barrow Neurological Institute in Phoenix describes the VMpo and the two insula regions as "a system that represents the material me."[4]

Among its other functions, the thalamus is considered a gateway for preprocessed sensory information to enter the neocortex. In addition to the tactile information flowing through the VMpo, processed information from the optic nerve (which, as noted above, has already been substantially transformed) is sent to a region of the thalamus called the lateral geniculate nucleus, which then sends it on to the V1 region of the neocortex. Information from the auditory sense is passed through the medial geniculate nucleus of the thalamus en route to the early auditory regions of the neocortex. All of our sensory data (except, apparently, for the olfactory system, which uses the olfactory bulb instead) passes through specific regions of the thalamus.

The most significant role of the thalamus, however, is its continual communication with the neocortex. The pattern recognizers in the neocortex send tentative results to the thalamus and receive responses principally using both excitatory and inhibitory reciprocal signals from layer VI of each recognizer. Keep in mind that these are not wireless messages, so that there needs to be an extraordinary amount of actual wiring (in the form of axons) running between all regions of the neocortex and the thalamus. Consider the vast amount of real estate (in terms of the physical mass of connections required) for the hundreds of millions of pattern recognizers in the neocortex to be constantly checking in with the thalamus.[5]

So what are the hundreds of millions of neocortical pattern recognizers talking to the thalamus about? It is apparently an important conversation, because profound damage to the main region of the thalamus bilaterally can lead to prolonged unconsciousness. A person with a damaged thalamus may still have activity in his neocortex, in that the self-triggering thinking by association can still work. But directed thinking—the kind that will get us out of bed, into our car, and sitting at our desk at work—does not function without a thalamus. In a famous case, twenty-one-year-old Karen Ann Quinlan suffered a heart attack and respiratory failure and remained in an unresponsive, apparently vegetative state for ten years. When she died, her autopsy revealed that her neocortex was normal but her thalamus had been destroyed.

In order to play its key role in our ability to direct attention, the thalamus relies on the structured knowledge contained in the neocortex. It can step through a list (stored in the neocortex), enabling us to follow a train of thought or follow a plan of action. We are apparently able to keep up to about four items in our working memory at a time, two per hemisphere according to recent research by neuroscientists at the MIT Picower Institute for Learning and Memory.[6] The issue of whether the thalamus is in charge of the neocortex or vice versa is far from clear, but we are unable to function without both.

The Hippocampus

Each brain hemisphere contains a hippocampus, a small region that looks like a sea horse tucked in the medial temporal lobe. Its primary function is to remember novel events. Since sensory information flows through the neocortex, it is up to the neocortex to determine that an experience is novel in order to present it to the hippocampus. It does so either by failing to recognize a particular set of features (for example, a new face) or by realizing that an otherwise familiar situation now has unique attributes (such as your spouse's wearing a fake mustache).

The hippocampus is capable of remembering these situations, although it appears to do so primarily through pointers into the neocortex. So memories in the hippocampus are also stored as lower-level patterns that were earlier recognized and stored in the neocortex. For animals without a neocortex to modulate sensory experiences, the hippocampus will simply remember the information from the senses, although this will have undergone sensory preprocessing (for example, the transformations performed by the optic nerve).

Although the hippocampus makes use of the neocortex (if a particular brain has one) as its scratch pad, its memory (of pointers into the neocortex) is not inherently hierarchical. Animals without a neocortex can accordingly remember things using their hippocampus, but their recollections will not be hierarchical.

The capacity of the hippocampus is limited, so its memory is short-term. It will transfer a particular sequence of patterns from its short-term memory to the long-term hierarchical memory of the neocortex by playing this memory sequence to the neocortex over and over again. We need, therefore, a hippocampus in order to learn new memories and skills (although strictly motor skills appear to use a different mechanism). Someone with damage to both copies of her hippocampus will retain her existing memories but will not be able to form new ones.

University of Southern California neuroscientist Theodore Berger and his colleagues modeled the hippocampus of a rat and have successfully experimented with implanting an artificial one. In a study reported in 2011, the USC scientists blocked particular learned behaviors in rats with drugs. Using an artificial hippocampus, the rats were able to quickly relearn the behavior. "Flip the switch on, and the rats remember. Flip it off and the rats forget," Berger wrote, referring to his ability to control the neural implant remotely. In another experiment the scientists allowed their artificial hippocampus to work alongside the rats' natural one. The result was that the ability of the rats to learn new behaviors strengthened. "These integrated experimental modeling studies show for the first time," Berger explained, "that . . . a neural prosthesis capable of real-time identification and manipulation of the encoding process can restore and even enhance cognitive mnemonic processes."[7] The hippocampus is one of the first regions damaged by Alzheimer's, so one goal of this research is to develop a neural implant for humans that will mitigate this first phase of damage from the disease.

The Cerebellum

There are two approaches you can use to catch a fly ball. You could solve the complex simultaneous differential equations controlling the ball's movement as well as further equations governing your own particular angle in viewing the ball, and then compute even more equations on how to move your body, arm, and hand to be in the right place at the right time.

This is not the approach that your brain adopts. It basically simplifies the problem by collapsing a lot of equations into a simple trend model, considering the trends of where the ball appears to be in your field of vision and how quickly it is moving within it. It does the same thing with your hand, making essentially linear predictions of the ball's apparent position in your field of view and that of your hand. The goal, of course, is to make sure they meet at the same point in space and time. If the ball appears to be dropping too quickly and your hand appears to be moving too slowly, your brain will direct your hand to move more quickly, so that the trends will coincide. This "Gordian knot" solution to what would otherwise be an intractable mathematical problem is called basis functions, and they are carried out by the cerebellum, a bean-shaped and appropriately baseball-sized region that sits on the brain stem.[8]

The cerebellum is an old-brain region that once controlled virtually all hominid movements. It still contains half of the neurons in the brain, although most are relatively small ones, so the region constitutes only about 10 percent of the weight of the brain. The cerebellum likewise represents another instance of massive repetition in the design of the brain. There is relatively little information about its design in the genome, as its structure is a pattern of several neurons that is repeated billions of times. As with the neocortex, there is uniformity across its structure.[9]

Most of the function of controlling our muscles has been taken over by the neocortex, using the same pattern recognition algorithms that it uses for perception and cognition. In the case of movement, we can more appropriately refer to the neocortex's function as pattern implementation. The

neocortex does make use of the memory in the cerebellum to record delicate scripts of movements—for example, your signature and certain flour-
ishes in artistic expression such as music and dance. Studies of the role of
the cerebellum during the learning of handwriting by children reveal that
the Purkinje cells of the cerebellum actually sample the sequence of movements, with each one sensitive to a specific sample.[10] Because most of our
movement is now controlled by the neocortex, many people can manage
with a relatively modest obvious disability even with significant damage to
the cerebellum, except that their movements may become less graceful.

The neocortex can also call upon the cerebellum to use its ability to
compute real-time basis functions to anticipate what the results of actions
would be that we are considering but have not yet carried out (and may
never carry out), as well as the actions or possible actions of others. It is
another example of the innate built-in linear predictors in the brain.

Substantial progress has been made in simulating the cerebellum with
respect to the ability to respond dynamically to sensory cues using the
basis functions I discussed above, in both bottom-up simulations (based
on biochemical models) and top-down simulations (based on mathematical models of how each repeating unit in the cerebellum operates).[11]

Pleasure and Fear

> Fear is the main source of superstition, and one of the main sources
> of cruelty. To conquer fear is the beginning of wisdom.
>
> —Bertrand Russell

> Feel the fear and do it anyway.
>
> —Susan Jeffers

If the neocortex is good at solving problems, then what is the main problem we are trying to solve? The problem that evolution has always tried to
solve is survival of the species. That translates into the survival of the

individual, and each of us uses his or her own neocortex to interpret that in myriad ways. In order to survive, animals need to procure their next meal while at the same time avoiding becoming someone else's meal. They also need to reproduce. The earliest brains evolved pleasure and fear systems that rewarded the fulfillment of these fundamental needs along with basic behaviors that facilitated them. As environments and competing species gradually changed, biological evolution made corresponding alterations. With the advent of hierarchical thinking, the satisfaction of critical drives became more complex, as it was now subject to the vast complex of ideas within ideas. But despite its considerable modulation by the neocortex, the old brain is still alive and well and still motivating us with pleasure and fear.

One region that is associated with pleasure is the nucleus accumbens. In famous experiments conducted in the 1950s, rats that were able to directly stimulate this small region (by pushing a lever that activated implanted electrodes) preferred doing so to anything else, including having sex or eating, ultimately exhausting and starving themselves to death.[12] In humans, other regions are also involved in pleasure, such as the ventral pallidum and, of course, the neocortex itself.

Pleasure is also regulated by chemicals such as dopamine and serotonin. It is beyond the scope of this book to discuss these systems in detail, but it is important to recognize that we have inherited these mechanisms from our premammalian cousins. It is the job of our neocortex to enable us to be the master of pleasure and fear and not their slave. To the extent that we are often subject to addictive behaviors, the neocortex is not always successful in this endeavor. Dopamine in particular is a neurotransmitter involved in the experience of pleasure. If anything good happens to us—winning the lottery, gaining the recognition of our peers, getting a hug from a loved one, or even subtle achievements such as getting a friend to laugh at a joke—we experience a release of dopamine. Sometimes we, like the rats who died overstimulating their nucleus accumbens, use a shortcut to achieve these bursts of pleasure, which is not always a good idea.

Gambling, for example, can release dopamine, at least when you win, but this is dependent on its inherent lack of predictability. Gambling may work for the purpose of releasing dopamine for a while, but given that the odds are intentionally stacked against you (otherwise the business model of a casino wouldn't work), it can become ruinous as a regular strategy. Similar dangers are associated with any addictive behavior. A particular genetic mutation of the dopamine-receptor D2 gene causes especially strong feelings of pleasure from initial experiences with addictive substances and behaviors, but as is well known (but not always well heeded), the ability of these substances to produce pleasure on subsequent use gradually declines. Another genetic mutation results in people's not receiving normal levels of dopamine release from everyday accomplishments, which can also lead to seeking enhanced early experiences with addictive activities. The minority of the population that has these genetic proclivities to addiction creates an enormous social and medical problem. Even those who manage to avoid severely addictive behaviors struggle with balancing the rewards of dopamine release with the consequences of the behaviors that release them.

Serotonin is a neurotransmitter that plays a major role in the regulation of mood. In higher levels it is associated with feelings of well-being and contentment. Serotonin has other functions, including modulating synaptic strength, appetite, sleep, sexual desire, and digestion. Antidepression drugs such as selective serotonin reuptake inhibitors (which tend to increase serotonin levels available to receptors) tend to have far-reaching effects, not all of them desirable (such as suppressing libido). Unlike actions in the neocortex, where recognition of patterns and activations of axons affect only a small number of neocortical circuits at a time, these substances affect large regions of the brain or even the entire nervous system.

Each hemisphere of the human brain has an amygdala, which consists of an almond-shaped region comprising several small lobes. The amygdala is also part of the old brain and is involved in processing a number of types of emotional responses, the most notable of which is fear. In premammalian animals, certain preprogrammed stimuli representing danger feed directly into the amygdala, which in turn triggers the "fight or flight" mechanism. In humans the amygdala now depends on perceptions of danger to be transmitted by the neocortex. A negative comment by your boss, for example, might trigger such a response by generating the fear of losing your job (or maybe not, if you have confidence in a plan B). Once the amygdala does decide that danger is ahead, an ancient sequence of events occurs. The amygdala signals the pituitary gland to release a hormone called ACTH (adrenocorticotropin). This in turn triggers the stress hormone cortisol from the adrenal glands, which results in more energy being provided to your muscles and nervous system. The adrenal glands also produce adrenaline and noradrenaline, which suppress your digestive, immune, and reproductive systems (figuring that these are not high-priority processes in an emergency). Levels of blood pressure, blood sugar, cholesterol, and fibrinogen (which speeds blood clotting) all rise. Heart rate and respiration go up. Even your pupils dilate so that you have better visual acuity of your enemy or your escape route. This is all very useful if a real danger such as a predator suddenly crosses your path. It is well known that in today's world, the chronic activation of this fight-or-flight mechanism can lead to permanent health damage in terms of hypertension, high cholesterol levels, and other problems.

The system of global neurotransmitter levels, such as serotonin, and hormone levels, such as dopamine, is intricate, and we could spend the rest of this book on the issue (as a great many books have done), but it is worth pointing out that the bandwidth of information (the rate of information processing) in this system is very low compared with the bandwidth of the neocortex. There are only a limited number of substances involved and the levels of these chemicals tend to change slowly and are relatively universal across the brain, as compared with the neocortex, which is composed of hundreds of trillions of connections that can change quickly.

It is fair to say that our emotional experiences take place in both the old and the new brains. Thinking takes place in the new brain (the neocortex), but feeling takes place in both. Any emulation of human behavior will therefore need to model both. However, if it is just human cognitive intelligence that we are after, the neocortex is sufficient. We can replace the old brain with the more direct motivation of a nonbiological neocortex to achieve the goals that we assign to it. For example, in the case of Watson, the goal was simply stated: Come up with correct answers to *Jeopardy!* queries (albeit these were further modulated by a program that understood *Jeopardy!* wagering). In the case of the new Watson system being jointly developed by Nuance and IBM for medical knowledge, the goal is to help treat human disease. Future systems can have goals such as actually curing disease and alleviating poverty. A lot of the pleasure-fear struggle is already obsolete for humans, as the old brain evolved long before even primitive human society got started; indeed most of it is reptilian.

There is a continual struggle in the human brain as to whether the old or the new brain is in charge. The old brain tries to set the agenda with its control of pleasure and fear experiences, whereas the new brain is continually trying to understand the relatively primitive algorithms of the old brain and seeking to manipulate it to its own agenda. Keep in mind that the amygdala is unable to evaluate danger on its own—in the human brain it relies on the neocortex to make those judgments. Is that person a friend or a foe, a lover or a threat? Only the neocortex can decide.

To the extent that we are not directly engaged in mortal combat and hunting for food, we have succeeded in at least partially sublimating our ancient drives to more creative endeavors. On that note, we'll discuss creativity and love in the next chapter.

CHAPTER 6

TRANSCENDENT ABILITIES

This is my simple religion. There is no need for temples; no need for complicated philosophy. Our own brain, our own heart is our temple; the philosophy is kindness.

—The Dalai Lama

My hand moves because certain forces—electric, magnetic, or whatever "nerve-force" may prove to be—are impressed on it by my brain. This nerve-force, stored in the brain, would probably be traceable, if Science were complete, to chemical forces supplied to the brain by the blood, and ultimately derived from the food I eat and the air I breathe.

—Lewis Carroll

Our emotional thoughts also take place in the neocortex but are influenced by portions of the brain ranging from ancient brain regions such as the amygdala to some evolutionarily recent brain structures such as the spindle neurons, which appear to play a key role in higher-level emotions. Unlike the regular and logical recursive structures found in the cerebral cortex, the spindle neurons have highly irregular shapes and connections. They are the largest neurons in the human brain, spanning its entire breadth. They are deeply interconnected, with hundreds of thousands of connections tying together diverse portions of the neocortex.

As mentioned earlier, the insula helps process sensory signals, but it also plays a key role in higher-level emotions. It is this region from which the spindle cells originate. Functional magnetic resonance imaging (fMRI) scans have revealed that these cells are particularly active when a person is dealing with emotions such as love, anger, sadness, and sexual desire. Situations that strongly activate them include when a subject looks at her partner or hears her child crying.

Spindle cells have long neural filaments called apical dendrites, which are able to connect to faraway neocortical regions. Such "deep" interconnectedness, in which certain neurons provide connections across numerous regions, is a feature that occurs increasingly as we go up the evolutionary ladder. It is not surprising that the spindle cells, involved as they are in handling emotion and moral judgment, would have this form of connectedness, given the ability of higher-level emotional reactions to touch on diverse topics and thoughts. Because of their links to many other parts of the brain, the high-level emotions that spindle cells process are affected by all of our perceptual and cognitive regions. It is important to point out that these cells are not doing rational problem solving, which is why we don't have rational control over our responses to music or over falling in love. The rest of the brain is heavily engaged, however, in trying to make sense of our mysterious high-level emotions.

There are relatively few spindle cells: only about 80,000, with approximately 45,000 in the right hemisphere and 35,000 in the left. This disparity is at least one reason for the perception that emotional intelligence is the province of the right brain, although the disproportion is modest. Gorillas have about 16,000 of these cells, bonobos about 2,100, and chimpanzees about 1,800. Other mammals lack them completely.

Anthropologists believe that spindle cells made their first appearance 10 to 15 million years ago in the as yet undiscovered common ancestor to apes and hominids (precursors to humans) and rapidly increased in numbers around 100,000 years ago. Interestingly, spindle cells do not exist in newborn humans but begin to appear only at around the age of four months and increase significantly in number from ages one to three. Children's ability to deal with moral issues and perceive such higher-level emotions as love develop during this same period.

Aptitude

Wolfgang Amadeus Mozart (1756–1791) wrote a minuet when he was five. At age six he performed for the empress Maria Theresa at the imperial court in Vienna. He went on to compose six hundred pieces, including forty-one symphonies, before his death at age thirty-five, and is widely regarded as the greatest composer in the European classical tradition. One might say that he had an aptitude for music.

So what does this mean in the context of the pattern recognition theory of mind? Clearly part of what we regard as aptitude is the product of nurture, that is to say, the influences of environment and other people. Mozart was born into a musical family. His father, Leopold, was a composer and kapellmeister (literally musical leader) of the court orchestra of the archbishop of Salzburg. The young Mozart was immersed in music, and his father started teaching him the violin and clavier (a keyboard instrument) at the age of three.

However, environmental influences alone do not fully explain Mozart's genius. There is clearly a nature component as well. What form does this take? As I wrote in chapter 4, different regions of the neocortex have become optimized (by biological evolution) for certain types of patterns. Even though the basic pattern recognition algorithm of the modules is uniform across the neocortex, since certain types of patterns tend to flow through particular regions (faces through the fusiform gyrus, for example), those regions will become better at processing the associated patterns. However, there are numerous parameters that govern how the algorithm is actually carried out in each module. For example, how close a match is required for a pattern to be recognized? How is that threshold modified if a higher-level module sends a signal that its pattern is "expected"? How are the size parameters considered? These and other factors have been set differently in different regions to be advantageous for particular types of patterns. In our work with similar methods in artificial intelligence, we have noticed the same phenomenon and have used simulations of evolution to optimize these parameters.

If particular regions can be optimized for different types of patterns, then it follows that individual brains will also vary in their ability to learn, recognize, and create certain types of patterns. For example, a brain can have an innate aptitude for music by being better able to recognize rhythmic patterns, or to better understand the geometric arrangements of harmonies. The phenomenon of perfect pitch (the ability to recognize and to reproduce a pitch without an external reference), which is correlated with musical talent, appears to have a genetic basis, although the ability needs to be developed, so it is likely to be a combination of nature and nurture. The genetic basis of perfect pitch is likely to reside outside the neocortex in the preprocessing of auditory information, whereas the learned aspect resides in the neocortex.

There are other skills that contribute to degrees of competency, whether of the routine variety or of the legendary genius. Neocortical abilities—for example, the ability of the neocortex to master the signals of fear that the amygdala generates (when presented with disapproval)—play a significant role, as do attributes such as confidence, organizational skills, and the ability to influence others. A very important skill I noted earlier is the courage to pursue ideas that go against the grain of orthodoxy. Invariably, people we regard as geniuses pursued their own mental experiments in ways that were not initially understood or appreciated by their peers. Although Mozart did gain recognition in his lifetime, most of the adulation came later. He died a pauper, buried in a common grave, and only two other musicians showed up at his funeral.

Creativity

> Creativity is a drug I cannot live without.
>
> —Cecil B. DeMille

> The problem is never how to get new, innovative thoughts into your mind, but how to get old ones out. Every mind is a building filled with archaic furniture. Clean out a corner of your mind and creativity will instantly fill it.
>
> —Dee Hock

> Humanity can be quite cold to those whose eyes see the world differently.
>
> —Eric A. Burns

> Creativity can solve almost any problem. The creative act, the defeat of habit by originality, overcomes everything.
>
> —George Lois

A key aspect of creativity is the process of finding great metaphors—symbols that represent something else. The neocortex is a great metaphor machine, which accounts for why we are a uniquely creative species. Every one of the approximately 300 million pattern recognizers in our neocortex is recognizing and defining a pattern and giving it a name, which in the case of the neocortical pattern recognition modules is simply the axon emerging from the pattern recognizer that will fire when that pattern is found. That symbol in turn then becomes part of another pattern. Each one of these patterns is essentially a metaphor. The recognizers can fire up to 100 times a second, so we have the potential of recognizing up to 30 billion metaphors a second. Of course not every module is firing in every cycle—but it is fair to say that we are indeed recognizing millions of metaphors a second.

Of course, some metaphors are more significant than others. Darwin perceived that Charles Lyell's insight on how very gradual changes from a trickle of water could carve out great canyons was a powerful metaphor for how a trickle of small evolutionary changes over thousands of generations could carve out great changes in the differentiation of species. Thought experiments, such as the one that Einstein used to illuminate the true meaning of the Michelson-Morley experiment, are all metaphors, in the sense of being a "thing regarded as representative or symbolic of something else," to quote a dictionary definition.

Do you see any metaphors in Sonnet 73 by Shakespeare?

That time of year thou mayst in me behold
When yellow leaves, or none, or few, do hang
Upon those boughs which shake against the cold,
Bare ruined choirs, where late the sweet birds sang.
In me thou seest the twilight of such day
As after sunset fadeth in the west,
Which by and by black night doth take away,
Death's second self that seals up all in rest.
In me thou seest the glowing of such fire
That on the ashes of his youth doth lie,
As the deathbed whereon it must expire
Consumed with that which it was nourished by.
This thou perceiv'st, which makes thy love more strong,
To love that well which thou must leave ere long.

In this sonnet, the poet uses extensive metaphors to describe his advancing age. His age is like late autumn, "when yellow leaves, or none, or few, do hang." The weather is cold and the birds can no longer sit on the branches, which he calls "bare ruin'd choirs." His age is like the twilight as the "sunset fadeth in the west, which by and by black night doth take away."

He is the remains of a fire "that on the ashes of his youth doth lie." Indeed, all language is ultimately metaphor, though some expressions of it are more memorable than others.

Finding a metaphor is the process of recognizing a pattern despite differences in detail and context—an activity we undertake trivially every moment of our lives. The metaphorical leaps that we consider of significance tend to take place in the interstices of different disciplines. Working against this essential force of creativity, however, is the pervasive trend toward ever greater specialization in the sciences (and just about every other field as well). As American mathematician Norbert Wiener (1894–1964) wrote in his seminal book *Cybernetics,* published the year I was born (1948):

> There are fields of scientific work, as we shall see in the body of this book, which have been explored from the different sides of pure mathematics, statistics, electrical engineering, and neurophysiology; in which every single notion receives a separate name from each group, and in which important work has been triplicated or quadruplicated, while still other important work is delayed by the unavailability in one field of results that may have already become classical in the next field.
>
> It is these boundary regions which offer the richest opportunities to the qualified investigator. They are at the same time the most refractory to the accepted techniques of mass attack and the division of labor.

A technique I have used in my own work to combat increasing specialization is to assemble the experts that I have gathered for a project (for example, my speech recognition work included speech scientists, linguists, psychoacousticians, and pattern recognition experts, not to mention computer scientists) and encourage each one to teach the group his particular techniques and terminology. We then throw out all of that terminology and make up our own. Invariably we find metaphors from one field that solve problems in another.

A mouse that finds an escape route when confronted with the household cat—and can do so even if the situation is somewhat different from what it has ever encountered before—is being creative. Our own creativity is orders of magnitude greater than that of the mouse—and involves far more levels of abstraction—because we have a much larger neocortex, which is capable of greater levels of hierarchy. So one way to achieve greater creativity is by effectively assembling more neocortex.

One approach to expand the available neocortex is through the collaboration of multiple humans. This is accomplished routinely via the communication between people gathered in a problem-solving community. Recently there have been efforts to use online collaboration tools to harness the power of real-time collaboration, which have shown success in mathematics and other fields.[1]

The next step, of course, will be to expand the neocortex itself with its nonbiological equivalent. This will be our ultimate act of creativity: to create the capability of being creative. A nonbiological neocortex will ultimately be faster and could rapidly search for the kinds of metaphors that inspired Darwin and Einstein. It could systematically explore all of the overlapping boundaries between our exponentially expanding frontiers of knowledge.

Some people express concern about what will happen to those who would opt out of such mind expansion. I would point out that this additional intelligence will essentially reside in the cloud (the exponentially expanding network of computers that we connect to through online communication), where most of our machine intelligence is now stored. When you use a search engine, recognize speech from your phone, consult a virtual assistant such as Siri, or use your phone to translate a sign into another language, the intelligence is not in the device itself but in the cloud. Our expanded neocortex will be housed there too. Whether we access such expanded intelligence through direct neural connection or the way we do now—by interacting with it via our devices—is an arbitrary distinction. In my view we will all become more creative through this pervasive enhancement, whether we choose to opt in or out of direct connection to humanity's expanded intelligence. We have already outsourced much of our personal, social, historical, and cultural memory to the cloud, and we will ultimately do the same thing with our hierarchical thinking.

Einstein's breakthrough resulted not only from his application of metaphors through mind experiments but also from his courage in believing in the power of those metaphors. He was willing to relinquish the traditional explanations that failed to satisfy his experiments, and he was willing to withstand the ridicule of his peers to the bizarre explanations that his metaphors implied. These qualities—belief in metaphor and courage of conviction—are ones that we should be able to program into our nonbiological neocortex as well.

Love

> Clarity of mind means clarity of passion, too; this is why a great and clear mind loves ardently and sees distinctly what it loves.
>
> —Blaise Pascal

> There is always some madness in love. But there is also always some reason in madness.
> —Friedrich Nietzsche

> When you have seen as much of life as I have, you will not underestimate the power of obsessive love.
>
> —Albus Dumbledore, in J. K. Rowling, *Harry Potter and the Half-Blood Prince*

> I always like a good math solution to any love problem.
>
> —Michael Patrick King, from the "Take Me Out to the Ballgame" episode of *Sex and the City*

If you haven't actually experienced ecstatic love personally, you have undoubtedly heard about it. It is fair to say that a substantial fraction if not a majority of the world's art—stories, novels, music, dance, paintings, television shows, and movies—is inspired by the stories of love in its earliest stages.

Science has recently gotten into the act as well, and we are now able to identify the biochemical changes that occur when someone falls in love. Dopamine is released, producing feelings of happiness and delight. Norepinephrine levels soar, which lead to a racing heart and overall feelings of exhilaration. These chemicals, along with phenylethylamine, produce elation, high energy levels, focused attention, loss of appetite, and a general craving for the object of one's desire. Interestingly, recent research at University College in London also shows that serotonin levels go down, similar to what happens in obsessive-compulsive disorder, which is consistent with the obsessive nature of early love.[2] The high levels of dopamine and norepinephrine account for the heightened short-term attention, euphoria, and craving of early love.

If these biochemical phenomena sound similar to those of the fight-or-flight syndrome, they are, except that here we are running toward something or someone; indeed, a cynic might say toward rather than away from danger. The changes are also fully consistent with those of the early phases of addictive behavior. The Roxy Music song "Love Is the Drug" is quite accurate in describing this state (albeit the subject of the song is looking to score his next fix of love). Studies of ecstatic religious experiences also show the same physical phenomena; it can be said that the person having such an experience is falling in love with God or whatever spiritual connection on which they are focused.

In the case of early romantic love, estrogen and testosterone certainly play a role in establishing sex drive, but if sexual reproduction were the only evolutionary objective of love, then the romantic aspect of the process would not be necessary. As psychologist John William Money (1921–2006) wrote, "Lust is lewd, love is lyrical."

The ecstatic phase of love leads to the attachment phase and ultimately to a long-term bond. There are chemicals that encourage this process as well, including oxytocin and vasopressin. Consider two related species of voles: the prairie vole and the montane vole. They are pretty much identical, except that the prairie vole has receptors for oxytocin and vasopressin, whereas the montane vole does not. The prairie vole is noted for lifetime monogamous relationships, while the montane vole resorts almost exclusively to one-night stands. In the case of voles, the oxytocin and vasopressin receptors are pretty much determinative as to the nature of their love life.

While these chemicals are influential on humans as well, our neocortex has taken a commanding role, as in everything else we do. Voles do have a neocortex, but it is postage-stamp sized and flat and just large enough for them to find a mate for life (or, in the case of montane voles, at least for the night) and carry out other basic vole behaviors. We humans have sufficient additional neocortex to engage in the expansive "lyrical" expressions to which Money refers.

From an evolutionary perspective, love itself exists to meet the needs of the neocortex. If we didn't have a neocortex, then lust would be quite sufficient to guarantee reproduction. The ecstatic instigation of love leads to attachment and mature love, and results in a lasting bond. This in turn is designed to provide at least the possibility of a stable environment for children while their own neocortices undergo the critical learning needed to become responsible and capable adults. Learning in a rich environment is inherently part of the method of the neocortex. Indeed the same oxytocin and vasopressin hormone mechanisms play a key role in establishing the critical bonding of parent (especially mother) and child.

At the far end of the story of love, a loved one becomes a major part of our neocortex. After decades of being together, a virtual other exists in the neocortex such that we can anticipate every step of what our lover will say and do. Our neocortical patterns are filled with the thoughts and patterns that reflect who they are. When we lose that person, we literally lose part of ourselves. This is not *just* a metaphor—all of the vast pattern recognizers that are filled with the patterns reflecting the person we love suddenly change their nature. Although they can be considered a precious way to keep that person alive within ourselves, the vast neocortical patterns of a lost loved one turn suddenly from triggers of delight to triggers of mourning.

The evolutionary basis for love and its phases is not the full story in today's world. We have already largely succeeded in liberating sex from its biological function, in that we can have babies without sex and we can certainly have sex without babies. The vast majority of sex takes place for its sensual and relational purposes. And we routinely fall in love for purposes other than raising children.

Similarly, the vast expanse of artistic expression of all kinds that celebrates love and its myriad forms dating back to antiquity is also an end in itself. Our ability to create these enduring forms of transcendent knowledge—about love or anything else—is precisely what makes our species unique.

The neocortex is biology's greatest creation. In turn, it is the poems about love—and all of our other creations—that represent the greatest inventions of our neocortex.

CHAPTER 7

THE BIOLOGICALLY INSPIRED
DIGITAL NEOCORTEX

Never trust anything that can think for itself if you can't see where it keeps its brain.

—Arthur Weasley, in J. K. Rowling,

Harry Potter and the Prisoner of Azkaban

No, I'm not interested in developing a powerful brain. All I'm after is just a mediocre brain, something like the President of the American Telephone and Telegraph Company.

—Alan Turing

A computer would deserve to be called intelligent if it could deceive a human into believing that it was human.

—Alan Turing

I believe that at the end of the century the use of words and general educated opinion will have altered so much that one will be able to speak of machines thinking without expecting to be contradicted.

—Alan Turing

A mother rat will build a nest for her young even if she has never seen another rat in her lifetime.[1] Similarly, a spider will spin a web, a caterpillar will create her own cocoon, and a beaver will build a dam, even if no contemporary ever showed them how to accomplish these complex tasks. That is not to say that these are not learned behaviors. It is just that these animals did not learn them in a single lifetime—they learned them over thousands of lifetimes. The evolution of animal behavior does constitute a learning process, but it is learning by the species, not by the individual, and the fruits of this learning process are encoded in DNA.

To appreciate the significance of the evolution of the neocortex, consider that it greatly sped up the process of learning (hierarchical knowledge) from thousands of years to months (or less). Even if millions of animals in a particular mammalian species failed to solve a problem (requiring a hierarchy of steps), it required only one to accidentally stumble upon a solution. That new method would then be copied and spread exponentially through the population.

We are now in a position to speed up the learning process by a factor of thousands or millions once again by migrating from biological to non-biological intelligence. Once a digital neocortex learns a skill, it can transfer that know-how in minutes or even seconds. As one of many examples, at my first company, Kurzweil Computer Products (now Nuance Speech Technologies), which I founded in 1973, we spent years training a set of research computers to recognize printed letters from scanned documents, a technology called omni-font (any type font) optical character recognition (OCR). This particular technology has now been in continual development for almost forty years, with the current product called OmniPage from Nuance. If you want your computer to recognize printed letters, you don't need to spend years training it to do so, as we did—you can simply download the evolved patterns already learned by the research computers in the form of software. In the 1980s we began on speech recognition, and that technology, which has also been in continuous development now for several decades, is part of Siri. Again, you can download in seconds the evolved patterns learned by the research computers over many years.

Ultimately we will create an artificial neocortex that has the full range and flexibility of its human counterpart. Consider the benefits. Electronic circuits are millions of times faster than our biological circuits. At first we will have to devote all of this speed increase to compensating for the relative lack of parallelism in our computers, but ultimately the digital neocortex will be much faster than the biological variety and will only continue to increase in speed.

When we augment our own neocortex with a synthetic version, we won't have to worry about how much additional neocortex can physically fit into our bodies and brains, as most of it will be in the cloud, like most of the computing we use today. I estimated earlier that we have on the order of 300 million pattern recognizers in our biological neocortex. That's as much as could be squeezed into our skulls even with the evolutionary innovation of a large forehead and with the neocortex taking about 80 percent of the available space. As soon as we start thinking in the cloud, there will be no natural limits—we will be able to use billions or trillions of pattern recognizers, basically whatever we need, and whatever the law of accelerating returns can provide at each point in time.

In order for a digital neocortex to learn a new skill, it will still require many iterations of education, just as a biological neocortex does, but once a single digital neocortex somewhere and at some time learns something, it can share that knowledge with every other digital neocortex without delay. We can each have our own private neocortex extenders in the cloud, just as we have our own private stores of personal data today.

Last but not least, we will be able to back up the digital portion of our intelligence. As we have seen, it is not just a metaphor to state that there is information contained in our neocortex, and it is frightening to contemplate that none of this information is backed up today. There is, of course, one way in which we do back up some of the information in our brains—by writing it down. The ability to transfer at least some of our thinking to a medium that can outlast our biological bodies was a huge step forward, but a great deal of data in our brains continues to remain vulnerable.

Brain Simulations

One approach to building a digital brain is to simulate precisely a biological one. For example, Harvard brain sciences doctoral student David Dalrymple (born in 1991) is planning to simulate the brain of a nematode (a roundworm).[2] Dalrymple selected the nematode because of its relatively simple nervous system, which consists of about 300 neurons, and which he plans to simulate at the very detailed level of molecules. He will also create a computer simulation of its body as well as its environment so that his virtual nematode can hunt for (virtual) food and do the other things that nematodes are good at. Dalrymple says it is likely to be the first complete brain upload from a biological animal to a virtual one that lives in a virtual world. Like his simulated nematode, whether even biological nematodes are conscious is open to debate, although in their struggle to eat, digest food, avoid predators, and reproduce, they do have experiences to be conscious of.

At the opposite end of the spectrum, Henry Markram's Blue Brain Project is planning to simulate the human brain, including the entire neocortex as well as the old-brain regions such as the hippocampus, amygdala, and cerebellum. His planned simulations will be built at varying degrees of detail, up to a full simulation at the molecular level. As I reported in chapter 4, Markram has discovered a key module of several dozen neurons that is repeated over and over again in the neocortex, demonstrating that learning is done by these modules and not by individual neurons.

Markram's progress has been scaling up at an exponential pace. He simulated one neuron in 2005, the year the project was initiated. In 2008 his team simulated an entire neocortical column of a rat brain, consisting of 10,000 neurons. By 2011 this expanded to 100 columns, totaling a million cells, which he calls a mesocircuit. One controversy concerning Markram's work is how to verify that the simulations are accurate. In order to do this, these simulations will need to demonstrate learning that I discuss below.

He projects simulating an entire rat brain of 100 mesocircuits, totaling 100 million neurons and about a trillion synapses, by 2014. In a talk at the 2009 TED conference at Oxford, Markram said, "It is not impossible to build a human brain, and we can do it in 10 years."[3] His most recent target for a full brain simulation is 2023.[4]

Markram and his team are basing their model on detailed anatomical and electrochemical analyses of actual neurons. Using an automated device they created called a patch-clamp robot, they are measuring the specific ion channels, neurotransmitters, and enzymes that are responsible for the electrochemical activity within each neuron. Their automated

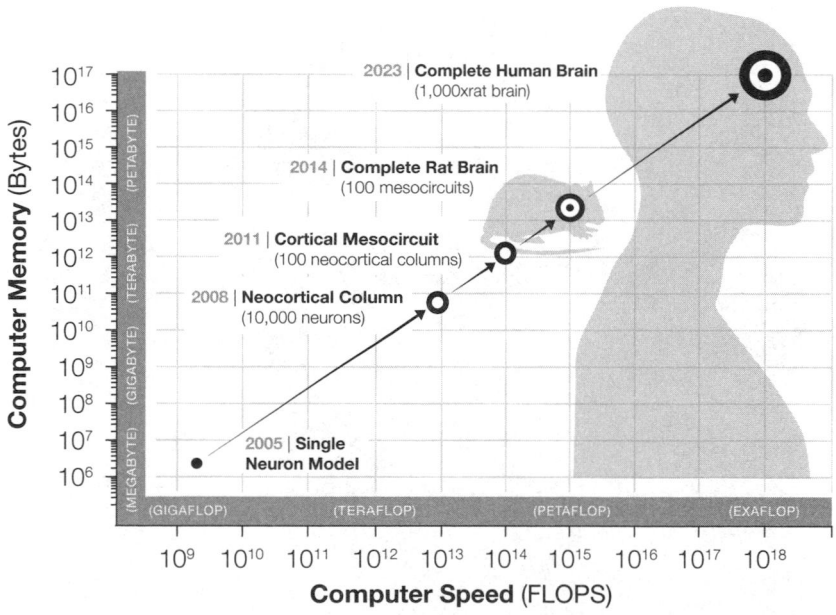

Actual and projected progress of the Blue Brain brain simulation project.

system was able to do thirty years of analysis in six months, according to Markram. It was from these analyses that they noticed the "Lego memory" units that are the basic functional units of the neocortex.

Significant contributions to the technology of robotic patch-clamping was made by MIT neuroscientist Ed Boyden, Georgia Tech mechanical engineering professor Craig Forest, and Forest's graduate student Suhasa Kodandaramaiah. They demonstrated an automated system with one-micrometer precision that can perform scanning of neural tissue at very close range without damaging the delicate membranes of the neurons. "This is something a robot can do that a human can't," Boyden commented.

To return to Markram's simulation, after simulating one neocortical column, Markram was quoted as saying, "Now we just have to scale it up."[5] Scaling is certainly one big factor, but there is one other key hurdle, which

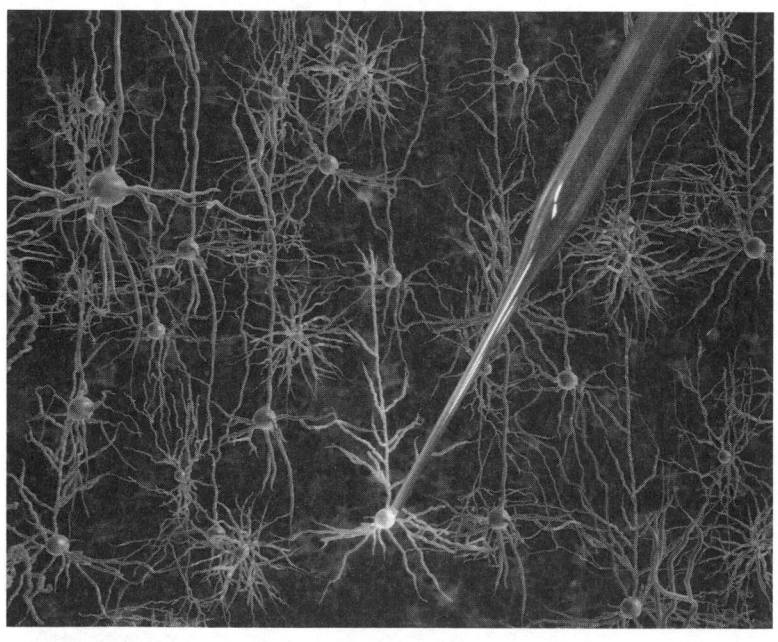

The tip of the patch-clamping robot developed at MIT and Georgia Tech scanning neural tissue.

is learning. If the Blue Brain Project brain is to "speak and have an intelligence and behave very much as a human does," which is how Markram described his goal in a BBC interview in 2009, then it will need to have sufficient content in its simulated neocortex to perform those tasks.[6] As anyone who has tried to hold a conversation with a newborn can attest, there is a lot of learning that must be achieved before this is feasible.

There are two obvious ways this can be done in a simulated brain such as Blue Brain. One would be to have the brain learn this content the way a human brain does. It can start out like a newborn human baby with an innate capacity for acquiring hierarchical knowledge and with certain transformations preprogrammed in its sensory preprocessing regions. But the learning that takes place between a biological infant and a human person who can hold a conversation would need to occur in a comparable manner in nonbiological learning. The problem with that approach is that a brain that is being simulated at the level of detail anticipated for Blue Brain is not expected to run in real time until at least the early 2020s. Even running in real time would be too slow unless the researchers are prepared to wait a decade or two to reach intellectual parity with an adult human, although real-time performance will get steadily faster as computers continue to grow in price/performance.

The other approach is to take one or more biological human brains that have already gained sufficient knowledge to converse in meaningful language and to otherwise behave in a mature manner and copy their neocortical patterns into the simulated brain. The problem with this method is that it requires a noninvasive and nondestructive scanning technology of sufficient spatial and temporal resolution and speed to perform such a task quickly and completely. I would not expect such an "uploading" technology to be available until around the 2040s. (The computational requirement to simulate a brain at that degree of precision, which I estimate to be 10^{19} calculations per second, will be available in a supercomputer according to my projections by the early 2020s; however, the necessary nondestructive brain scanning technologies will take longer.)

There is a third approach, which is the one I believe simulation projects such as Blue Brain will need to pursue. One can simplify molecular models by creating functional equivalents at different levels of specificity, ranging from my own functional algorithmic method (as described in this book) to simulations that are closer to full molecular simulations. The speed of learning can thereby be increased by a factor of hundreds or thousands depending on the degree of simplification used. An educational program can be devised for the simulated brain (using the functional model) that it can learn relatively quickly. Then the full molecular simulation can be substituted for the simplified model while still using its accumulated learning. We can then simulate learning with the full molecular model at a much slower speed.

American computer scientist Dharmendra Modha and his IBM colleagues have created a cell-by-cell simulation of a portion of the human visual neocortex comprising 1.6 billion virtual neurons and 9 trillion synapses, which is equivalent to a cat neocortex. It runs 100 times slower than real time on an IBM BlueGene/P supercomputer consisting of 147,456 processors. The work received the Gordon Bell Prize from the Association for Computing Machinery.

The purpose of a brain simulation project such as Blue Brain and Modha's neocortex simulations is specifically to refine and confirm a functional model. AI at the human level will principally use the type of functional algorithmic model discussed in this book. However, molecular simulations will help us to perfect that model and to fully understand which details are important. In my development of speech recognition technology in the 1980s and 1990s, we were able to refine our algorithms once the actual transformations performed by the auditory nerve and early portions of the auditory cortex were understood. Even if our functional model was perfect, understanding exactly how it is actually implemented in our biological brains will reveal important knowledge about human function and dysfunction.

We will need detailed data on actual brains to create biologically based simulations. Markram's team is collecting its own data. There are large-scale projects to gather this type of data and make it generally available to scientists. For example, Cold Spring Harbor Laboratory in New York has collected 500 terabytes of data by scanning a mammal brain (a mouse), which they made available in June 2012. Their project allows a user to explore a brain similarly to the way Google Earth allows one to explore the surface of the planet. You can move around the entire brain and zoom in to see individual neurons and their connections. You can highlight a single connection and then follow its path through the brain.

Sixteen sections of the National Institutes of Health have gotten together and sponsored a major initiative called the Human Connectome Project with $38.5 million of funding.[7] Led by Washington University in St. Louis, the University of Minnesota, Harvard University, Massachusetts General Hospital, and the University of California at Los Angeles, the project seeks to create a similar three-dimensional map of connections in the human brain. The project is using a variety of noninvasive scanning technologies, including new forms of MRI, magnetoencephalography (measuring the magnetic fields produced by the electrical activity in the brain), and diffusion tractography (a method to trace the pathways of fiber bundles in the brain). As I point out in chapter 10, the spatial resolution of noninvasive scanning of the brain is improving at an exponential rate. The research by Van J. Wedeen and his colleagues at Massachusetts General Hospital showing a highly regular gridlike structure of the wiring of the neocortex that I described in chapter 4 is one early result from this project.

Oxford University computational neuroscientist Anders Sandberg (born in 1972) and Swedish philosopher Nick Bostrom (born in 1973) have written the comprehensive *Whole Brain Emulation: A Roadmap,* which details the requirements for simulating the human brain (and other types of brains) at different levels of specificity from high-level functional models to simulating molecules.[8] The report does not provide a timeline, but it does describe the requirements to simulate different types of brains at varying levels of precision in terms of brain scanning, modeling, storage, and computation. The report projects ongoing exponential gains in all of these areas of capability and argues that the requirements to simulate the human brain at a high level of detail are coming into place.

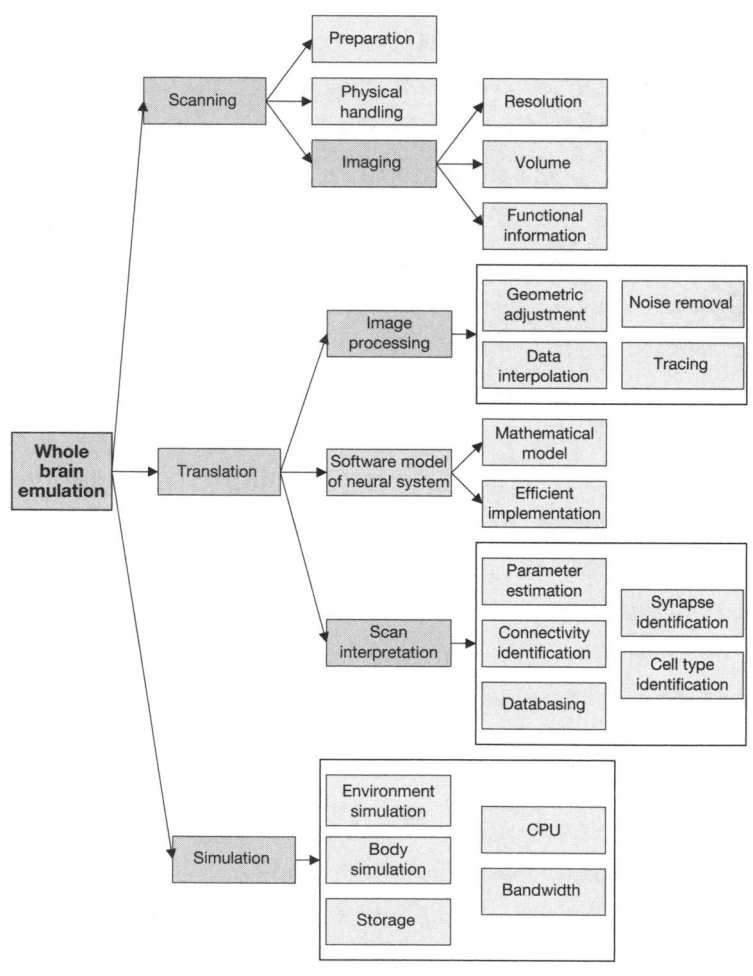

An outline of the technological capabilities needed for whole brain emulation, in *Whole Brain Emulation: A Roadmap* by Anders Sandberg and Nick Bostrom.

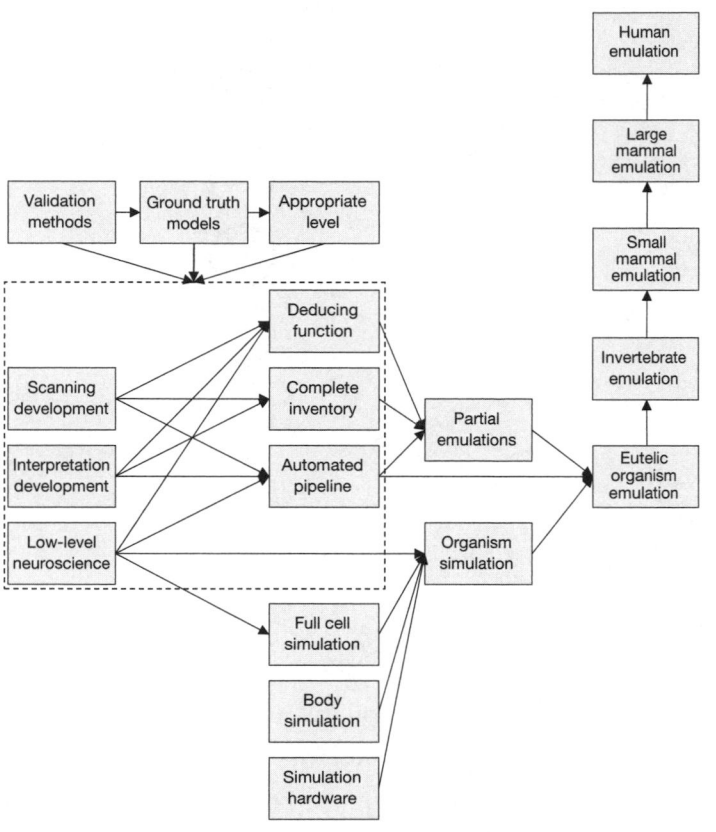

An outline of *Whole Brain Emulation: A Roadmap* by Anders Sandberg and Nick Bostrom.

Neural Nets

In 1964, at the age of sixteen, I wrote to Frank Rosenblatt (1928–1971), a professor at Cornell University, inquiring about a machine called the Mark 1 Perceptron. He had created it four years earlier, and it was described as having brainlike properties. He invited me to visit him and try the machine out.

The Perceptron was built from what he claimed were electronic models of neurons. Input consisted of values arranged in two dimensions. For speech, one dimension represented frequency and the other time, so each value represented the intensity of a frequency at a given point in time. For

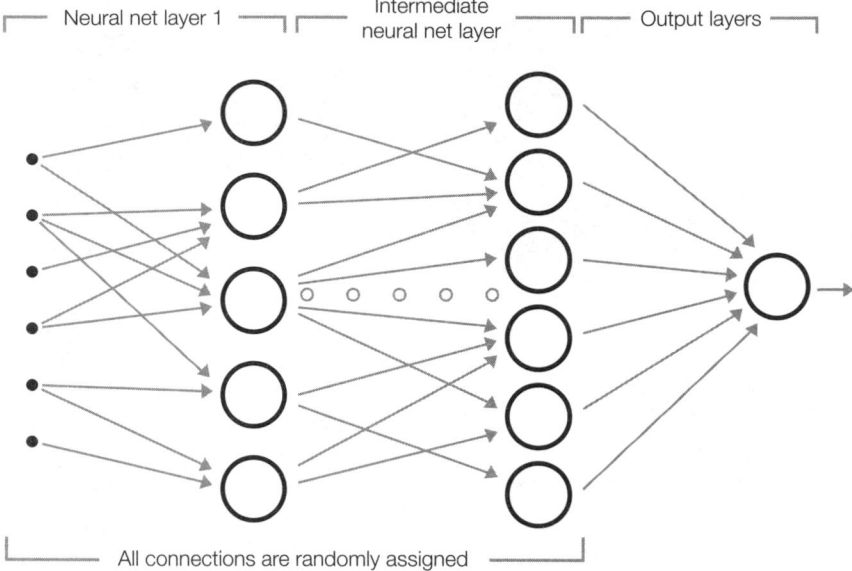

Neural net layer 1 — Intermediate neural net layer — Output layers

All connections are randomly assigned

images, each point was a pixel in a two-dimensional image. Each point of a given input was randomly connected to the inputs of the first layer of simulated neurons. Every connection had an associated synaptic strength, which represented its importance, and which was initially set at a random value. Each neuron added up the signals coming into it. If the combined signal exceeded a particular threshold, the neuron fired and sent a signal to its output connection; if the combined input signal did not exceed the threshold, the neuron did not fire, and its output was zero. The output of each neuron was randomly connected to the inputs of the neurons in the next layer. The Mark 1 Perceptron had three layers, which could be organized in a variety of configurations. For example, one layer might feed back to an earlier one. At the top layer, the output of one or more neurons, also randomly selected, provided the answer. (For an algorithmic description of neural nets, see this endnote.)[9]

Since the neural net wiring and synaptic weights are initially set randomly, the answers of an untrained neural net are also random. The key to a neural net, therefore, is that it must learn its subject matter, just like the mammalian brains on which it's supposedly modeled. A neural net starts out ignorant; its teacher—which may be a human, a computer program, or perhaps another, more mature neural net that has already learned its lessons—rewards the student neural net when it generates the correct output and punishes it when it does not. This feedback is in turn used by the student neural net to adjust the strength of each interneuronal connection. Connections that are consistent with the correct answer are made stronger. Those that advocate a wrong answer are weakened.

Over time the neural net organizes itself to provide the correct answers without coaching. Experiments have shown that neural nets can learn their subject matter even with unreliable teachers. If the teacher is correct only 60 percent of the time, the student neural net will still learn its lessons with an accuracy approaching 100 percent.

However, limitations in the range of material that the Perceptron was capable of learning quickly became apparent. When I visited Professor Rosenblatt in 1964, I tried simple modifications to the input. The system was set up to recognize printed letters, and would recognize them quite accurately. It did a fairly good job of autoassociation (that is, it could recognize the letters even if I covered parts of them), but fared less well with invariance (that is, generalizing over size and font changes, which confused it).

During the last half of the 1960s, these neural nets became enormously popular, and the field of "connectionism" took over at least half of the artificial intelligence field. The more traditional approach to AI, meanwhile, included direct attempts to program solutions to specific problems, such as how to recognize the invariant properties of printed letters.

Another person I visited in 1964 was Marvin Minsky (born in 1927), one of the founders of the artificial intelligence field. Despite having done some pioneering work on neural nets himself in the 1950s, he was concerned with the great surge of interest in this technique. Part of the allure of neural nets was that they supposedly did not require programming— they would learn solutions to problems on their own. In 1965 I entered MIT as a student with Professor Minsky as my mentor, and I shared his skepticism about the craze for "connectionism."

In 1969 Minsky and Seymour Papert (born in 1928), the two cofounders of the MIT Artificial Intelligence Laboratory, wrote a book called *Perceptrons*, which presented a single core theorem: specifically, that a Perceptron was inherently incapable of determining whether or not an image was connected. The book created a firestorm. Determining whether or not an image is connected is a task that humans can do very easily, and it is also a straightforward process to program a computer to make this

Two images from the cover of the book *Perceptrons* by Marvin Minsky and Seymour Papert. The top image is not connected (that is, the dark area consists of two disconnected parts). The bottom image is connected. A human can readily determine this, as can a simple software program. A feedforward Perceptron such as Frank Rosenblatt's Mark 1 Perceptron cannot make this determination.

discrimination. The fact that Perceptrons could not do so was considered by many to be a fatal flaw.

Perceptrons, however, was widely interpreted to imply more than it actually did. Minsky and Papert's theorem applied only to a particular type of neural net called a feedforward neural net (a category that does include Rosenblatt's Perceptron); other types of neural nets did not have this limitation. Still, the book did manage to largely kill most funding for neural net research during the 1970s. The field did return in the 1980s with attempts to use what were claimed to be more realistic models of biological neurons and ones that avoided the limitations implied by the Minsky-Papert Perceptron theorem. Nevertheless, the ability of the neocortex to solve the invariance problem, a key to its strength, was a skill that remained elusive for the resurgent connectionist field.

Sparse Coding: Vector Quantization

In the early 1980s I started a project devoted to another classical pattern recognition problem: understanding human speech. At first, we used traditional AI approaches by directly programming expert knowledge about the fundamental units of speech—phonemes—and rules from linguists on how people string phonemes together to form words and phrases. Each phoneme has distinctive frequency patterns. For example, we knew that vowels such as "e" and "ah" are characterized by certain resonant frequencies called formants, with a characteristic ratio of formants for each phoneme. Sibilant sounds such as "z" and "s" are characterized by a burst of noise that spans many frequencies.

We captured speech as a waveform, which we then converted into multiple frequency bands (perceived as pitches) using a bank of frequency filters. The result of this transformation could be visualized and was called a spectrogram (see page 136).

The filter bank is copying what the human cochlea does, which is the initial step in our biological processing of sound. The software first

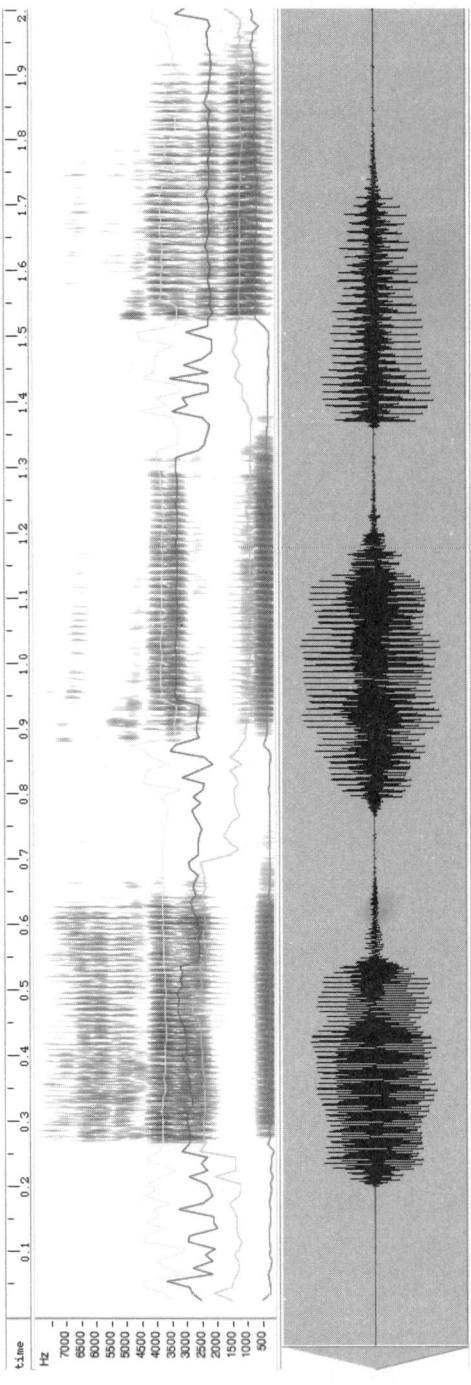

A spectrogram of three vowels. From left to right: [i] as in "appreciate," [u] as in "acoustic," and [a] as in "ah." The Y axis represents frequency of sound. The darker the band the more acoustic energy there is at that frequency.

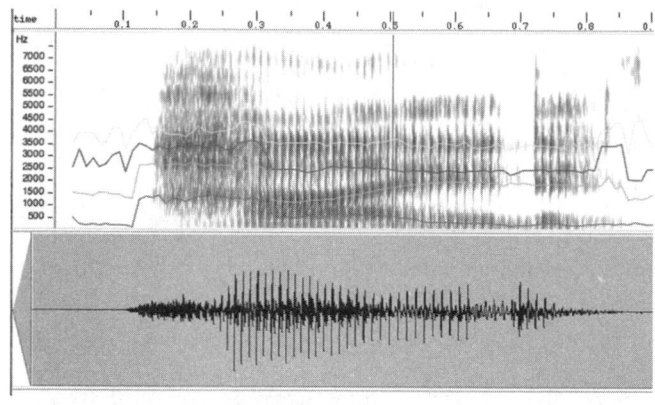

A spectrogram of a person saying the word "hide." The horizontal lines show the formants, which are sustained frequencies that have especially high energy.[10]

identified phonemes based on distinguishing patterns of frequencies and then identified words based on identifying characteristic sequences of phonemes.

The result was partially successful. We could train our device to learn the patterns for a particular person using a moderate-sized vocabulary, measured in thousands of words. When we attempted to recognize tens of thousands of words, handle multiple speakers, and allow fully continuous speech (that is, speech with no pauses between words), we ran into the invariance problem. Different people enunciated the same phoneme differently—for example, one person's "e" phoneme may sound like someone else's "ah." Even the same person was inconsistent in the way she spoke a particular phoneme. The pattern of a phoneme was often affected by other phonemes nearby. Many phonemes were left out completely. The pronunciation of words (that is, how phonemes are strung together to form words) was also highly variable and dependent on context. The linguistic rules we had programmed were breaking down and could not keep up with the extreme variability of spoken language.

It became clear to me at the time that the essence of human pattern and conceptual recognition was based on hierarchies. This is certainly apparent for human language, which constitutes an elaborate hierarchy of structures. But what is the element at the base of the structures? That was the first question I considered as I looked for ways to automatically recognize fully normal human speech.

Sound enters the ear as a vibration of the air and is converted by the approximately 3,000 inner hair cells in the cochlea into multiple frequency bands. Each hair cell is tuned to a particular frequency (note that we perceive frequencies as tones) and each acts as a frequency filter, emitting a signal whenever there is sound at or near its resonant frequency. As it leaves the human cochlea, sound is thereby represented by approximately 3,000 separate signals, each one signifying the time-varying intensity of a narrow band of frequencies (with substantial overlap among these bands).

Even though it was apparent that the brain was massively parallel, it seemed impossible to me that it was doing pattern matching on 3,000 separate auditory signals. I doubted that evolution could have been that inefficient. We now know that very substantial data reduction does indeed take place in the auditory nerve before sound signals ever reach the neocortex.

In our software-based speech recognizers, we also used filters implemented as software—sixteen to be exact (which we later increased to thirty-two, as we found there was not much benefit to going much higher than this). So in our system, each point in time was represented by sixteen numbers. We needed to reduce these sixteen streams of data into one while at the same emphasizing the features that are significant in recognizing speech.

We used a mathematically optimal technique to accomplish this, called vector quantization. Consider that at any particular point in time, sound (at least from one ear) was represented by our software by sixteen different numbers: that is, the output of the sixteen frequency filters. (In the human auditory system the figure would be 3,000, representing the output of the 3,000 cochlea inner hair cells.) In mathematical terminology, each such set of numbers (whether 3,000 in the biological case or 16 in our software implementation) is called a vector.

For simplicity, let's consider the process of vector quantization with vectors of two numbers. Each vector can be considered a point in two-dimensional space.

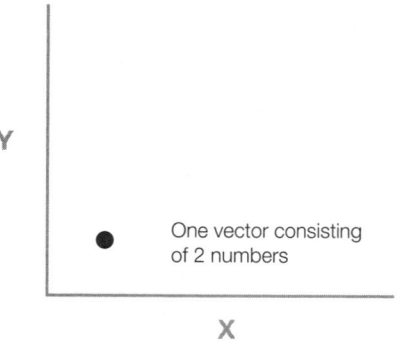

If we have a very large sample of such vectors and plot them, we are likely to notice clusters forming.

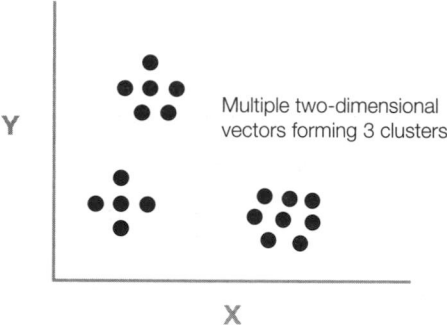

In order to identify the clusters, we need to decide how many we will allow. In our project we generally allowed 1,024 clusters so that we could number them and assign each cluster a 10-bit label (because 2^{10} = 1,024). Our sample of vectors represents the diversity that we expect. We tentatively assign the first 1,024 vectors to be one-point clusters. We then consider the 1,025th vector and find the point that it is closest to. If that distance is greater than the smallest distance between any pair of the 1,024 points, we consider it as the beginning of a new cluster. We then collapse the two (one-point) clusters that are closest together into a single cluster. We are thus still left with 1,024 clusters. After processing the 1,025th vector, one of those clusters now has more than one point. We keep processing points in this way, always maintaining 1,024 clusters. After we have processed all the points, we represent each multipoint cluster by the geometric center of the points in that cluster.

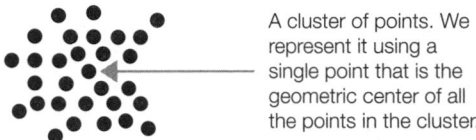

A cluster of points. We represent it using a single point that is the geometric center of all the points in the cluster.

We continue this iterative process until we have run through all the sample points. Typically we would process millions of points into 1,024 (2^{10}) clusters; we've also used 2,048 (2^{11}) or 4,096 (2^{12}) clusters. Each cluster is represented by one vector that is at the geometric center of all the points in that cluster. Thus the total of the distances of all the points in the cluster to the center point of the cluster is as small as possible.

The result of this technique is that instead of having the millions of points that we started with (and an even larger number of possible points), we have now reduced the data to just 1,024 points that use the space of possibilities optimally. Parts of the space that are never used are not assigned any clusters.

We then assign a number to each cluster (in our case, 0 to 1,023). That number is the reduced, "quantized" representation of that cluster, which is why the technique is called vector quantization. Any new input vector that arrives in the future is then represented by the number of the cluster whose center point is closest to this new input vector.

We can now precompute a table with the distance of the center point of every cluster to every other center point. We thereby have instantly available the distance of this new input vector (which we represent by this quantized point—in other words, by the number of the cluster that this new point is closest to) to every other cluster. Since we are only representing points by their closest cluster, we now know the distance of this point to any other possible point that might come along.

I described the technique above using vectors with only two numbers each, but working with sixteen-element vectors is entirely analogous to the simpler example. Because we chose vectors with sixteen numbers representing sixteen different frequency bands, each point in our system was a point in sixteen-dimensional space. It is difficult for us to imagine a space with more than three dimensions (perhaps four, if we include time), but mathematics has no such inhibitions.

We have accomplished four things with this process. First, we have greatly reduced the complexity of the data. Second, we have reduced sixteen-dimensional data to one-dimensional data (that is, each sample is now a single number). Third, we have improved our ability to find invariant features, because we are emphasizing portions of the space of possible sounds that convey the most information. Most combinations of frequencies are physically impossible or at least very unlikely, so there is no reason to give equal space to unlikely combinations of inputs as to likely ones. This technique reduces the data to equally likely possibilities. The fourth benefit is that we can use one-dimensional pattern recognizers, even though the original data consisted of many more dimensions. This turned out to be the most efficient approach to utilizing available computational resources.

Reading Your Mind with Hidden Markov Models

With vector quantization, we simplified the data in a way that emphasized key features, but we still needed a way to represent the hierarchy of invariant features that would make sense of new information. Having worked in the field of pattern recognition at that time (the early 1980s) for twenty years, I knew that one-dimensional representations were far more powerful, efficient, and amenable to invariant results. There was not a lot known about the neocortex in the early 1980s, but based on my experience with a variety of pattern recognition problems, I assumed that the brain was also likely to be reducing its multidimensional data (whether from the eyes, the ears, or the skin) using a one-dimensional representation, especially as concepts rose in the neocortex's hierarchy.

For the speech recognition problem, the organization of information in the speech signal appeared to be a hierarchy of patterns, with each pattern represented by a linear string of elements with a forward direction. Each element of a pattern could be another pattern at a lower level, or a fundamental unit of input (which in the case of speech recognition would be our quantized vectors).

You will recognize this situation as consistent with the model of the neocortex that I presented earlier. Human speech, therefore, is produced by a hierarchy of linear patterns in the brain. If we could simply examine these patterns in the brain of the person speaking, it would be a simple matter to match her new speech utterances against her brain patterns and understand what the person was saying. Unfortunately we do not have direct access to the brain of the speaker—the only information we have is what she actually said. Of course, that is the whole point of spoken language—the speaker is sharing a piece of her mind with her utterance.

So I wondered: Was there a mathematical technique that would enable us to infer the patterns in the speaker's brain based on her spoken words? One utterance would obviously not be sufficient, but if we had a large number of samples, could we use that information to essentially read the patterns in the speaker's neocortex (or at least formulate something mathematically equivalent that would enable us to recognize new utterances)?

People often fail to appreciate how powerful mathematics can be— keep in mind that our ability to search much of human knowledge in a fraction of a second with search engines is based on a mathematical technique. For the speech recognition problem I was facing in the early 1980s, it turned out that the technique of hidden Markov models fit the bill

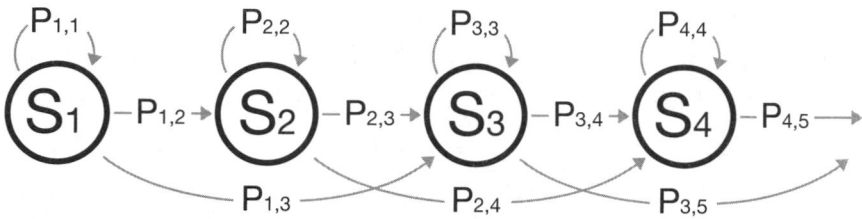

A simple example of one layer of a hidden Markov model. S_1 through S_4 represent the "hidden" internal states. The $P_{i,j}$ transitions each represent the probability of going from state S_i to state S_j. These probabilities are determined by the system learning from training data (including during actual use). A new sequence (such as a new spoken utterance) is matched against these probabilities to determine the likelihood that this model produced the sequence.

rather perfectly. The Russian mathematician Andrei Andreyevich Markov (1856–1922) built a mathematical theory of hierarchical sequences of states. The model was based on the possibility of traversing the states in one chain, and if that was successful, triggering a state in the next higher level in the hierarchy. Sound familiar?

Markov's model included probabilities of each state's successfully occurring. He went on to hypothesize a situation in which a system has such a hierarchy of linear sequences of states, but those are unable to be directly examined—hence the name *hidden* Markov models. The lowest level of the hierarchy emits signals, which are all we are allowed to see. Markov provides a mathematical technique to compute what the probabilities of each transition must be based on the observed output. The method was subsequently refined by Norbert Wiener in 1923. Wiener's refinement also provided a way to determine the connections in the Markov model; essentially any connection with too low a probability was considered not to exist. This is essentially how the human neocortex trims connections—if they are rarely or never used, they are considered unlikely and are pruned away. In our case, the observed output is the speech signal created by the person talking, and the state probabilities and connections of the Markov model constitute the neocortical hierarchy that produced it.

I envisioned a system in which we would take samples of human speech, apply the hidden Markov model technique to infer a hierarchy of states with connections and probabilities (essentially a simulated neocortex for producing speech), and then use this inferred hierarchical network of states to recognize new utterances. To create a speaker-independent system, we would use samples from many different individuals to train the hidden Markov models. By adding in the element of hierarchies to represent the hierarchical nature of information in language, these were properly called hierarchical hidden Markov models (HHMMs).

My colleagues at Kurzweil Applied Intelligence were skeptical that this technique would work, given that it was a self-organizing method reminiscent of neural nets, which had fallen out of favor and with which we had had little success. I pointed out that the network in a neural net system is fixed and does not adapt to the input: The weights adapt, but the connections do not. In the Markov model system, if it was set up correctly, the system would prune unused connections so as to essentially adapt the topology.

I established what was considered a "skunk works" project (an organizational term for a project off the beaten path that has little in the way of formal resources) that consisted of me, one part-time programmer, and an electrical engineer (to create the frequency filter bank). To the surprise of my colleagues, our effort turned out to be very successful, having succeeded in recognizing speech comprising a large vocabulary with high accuracy.

After that experiment, all of our subsequent speech recognition efforts have been based on hierarchical hidden Markov models. Other speech recognition companies appeared to discover the value of this method independently, and since the mid-1980s most work in automated speech recognition has been based on this approach. Hidden Markov models are also used in speech synthesis—keep in mind that our biological cortical hierarchy is used not only to recognize input but also to produce output, for example, speech and physical movement.

HHMMs are also used in systems that understand the meaning of

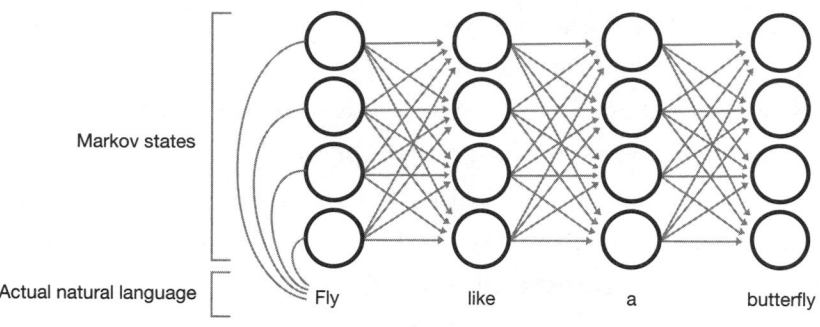

Hidden Markov states and possible transitions to produce a sequence of words in natural-language text.

natural-language sentences, which represents going up the conceptual hierarchy.

To understand how the HHMM method works, we start out with a network that consists of all the state transitions that are possible. The vector quantization method described above is critical here, because otherwise there would be too many possibilities to consider.

Here is a possible simplified initial topology:

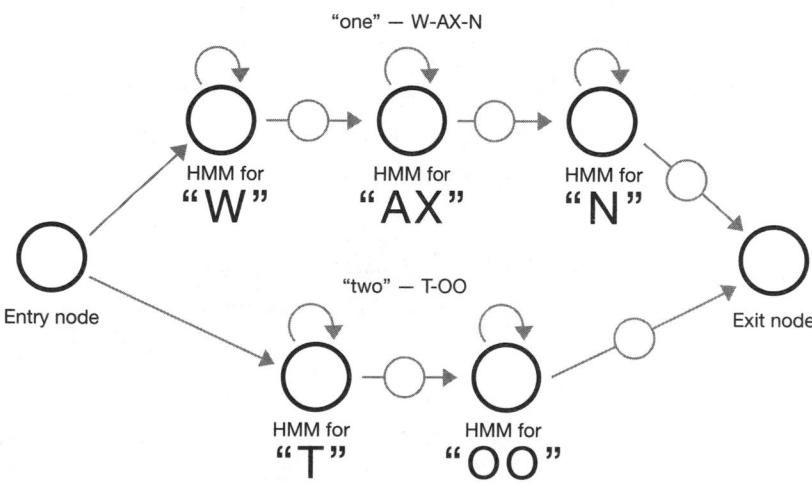

A simple hidden Markov model topology to recognize two spoken words.

Sample utterances are processed one by one. For each, we iteratively modify the probabilities of the transitions to better reflect the input sample we have just processed. The Markov models used in speech recognition code the likelihood that specific patterns of sound are found in each phoneme, how the phonemes influence one another, and the likely orders of phonemes. The system can also include probability networks on higher levels of language structure, such as the order of words, the inclusion of phrases, and so on up the hierarchy of language.

Whereas our previous speech recognition systems incorporated specific rules about phoneme structures and sequences explicitly coded by human linguists, the new HHMM-based system was not explicitly told that there are forty-four phonemes in English, the sequences of vectors that were likely for each phoneme, or what phoneme sequences were more likely than others. We let the system discover these "rules" for itself from thousands of hours of transcribed human speech data. The advantage of this approach over hand-coded rules is that the models develop probabilistic rules of which human experts are often not aware. We noticed that many of the rules that the system had automatically learned from the data differed in subtle but important ways from the rules established by human experts.

Once the network was trained, we began to attempt to recognize speech by considering the alternate paths through the network and picking the path that was most likely, given the actual sequence of input vectors we had seen. In other words, if we saw a sequence of states that was likely to have produced that utterance, we concluded that the utterance came from that cortical sequence. This simulated HHMM-based neocortex included word labels, so it was able to propose a transcription of what it heard.

We were then able to improve our results further by continuing to train the network while we were using it for recognition. As we have discussed, simultaneous recognition and learning also take place at every level in our biological neocortical hierarchy.

Evolutionary (Genetic) Algorithms

There is another important consideration: How do we set the many parameters that control a pattern recognition system's functioning? These could include the number of vectors that we allow in the vector quantization step, the initial topology of hierarchical states (before the training phase of the hidden Markov model process prunes them back), the recognition threshold at each level of the hierarchy, the parameters that control the handling of the size parameters, and many others. We can establish these based on our intuition, but the results will be far from optimal.

We call these parameters "God parameters" because they are set prior to the self-organizing method of determining the topology of the hidden Markov models (or, in the biological case, before the person learns her lessons by similarly creating connections in her cortical hierarchy). This is perhaps a misnomer, given that these initial DNA-based design details are determined by biological evolution, though some may see the hand of God in that process (and while I do consider evolution to be a spiritual process, this discussion properly belongs in chapter 9).

When it came to setting these "God parameters" in our simulated hierarchical learning and recognizing system, we again took a cue from nature and decided to evolve them—in our case, using a simulation of evolution. We used what are called genetic or evolutionary algorithms (GAs), which include simulated sexual reproduction and mutations.

Here is a simplified description of how this method works. First, we determine a way to code possible solutions to a given problem. If the problem is optimizing the design parameters for a circuit, then we define a list of all of the parameters (with a specific number of bits assigned to each parameter) that characterize the circuit. This list is regarded as the genetic code in the genetic algorithm. Then we randomly generate thousands or more genetic codes. Each such genetic code (which represents one set of design parameters) is considered a simulated "solution" organism.

Now we evaluate each simulated organism in a simulated environment by using a defined method to assess each set of parameters. This evaluation is a key to the success of a genetic algorithm. In our example, we would run each program generated by the parameters and judge it on appropriate criteria (did it complete the task, how long did it take, and so on). The best-solution organisms (the best designs) are allowed to survive, and the rest are eliminated.

Now we cause each of the survivors to multiply themselves until they reach the same number of solution creatures. This is done by simulating sexual reproduction: In other words, we create new offspring where each new creature draws one part of its genetic code from one parent and another part from a second parent. Usually no distinction is made between male or female organisms; it's sufficient to generate an offspring from any two arbitrary parents, so we're basically talking about same-sex marriage here. This is perhaps not as interesting as sexual reproduction in the natural world, but the relevant point here is having two parents. As these simulated organisms multiply, we allow some mutation (random change) in the chromosomes to occur.

We've now defined one generation of simulated evolution; now we repeat these steps for each subsequent generation. At the end of each generation we determine how much the designs have improved (that is, we compute the average improvement in the evaluation function over all the surviving organisms). When the degree of improvement in the evaluation of the design creatures from one generation to the next becomes very small, we stop this iterative cycle and use the best design(s) in the last generation. (For an algorithmic description of genetic algorithms, see this endnote.)[11]

The key to a genetic algorithm is that the human designers don't directly program a solution; rather, we let one emerge through an iterative process of simulated competition and improvement. Biological evolution is smart but slow, so to enhance its intelligence we greatly speed up its ponderous pace. The computer is fast enough to simulate many generations in a matter of hours or days, and we've occasionally had them run for as long as weeks to simulate hundreds of thousands of generations. But we have to go through this iterative process only once; as soon as we have let this simulated evolution run its course, we can apply the evolved and highly refined rules to real problems in a rapid fashion. In the case of our speech recognition systems, we used them to evolve the initial topology of the network and other critical parameters. We thus used two self-organizing methods: a GA to simulate the biological evolution that gave rise to a particular cortical design, and HHMMs to simulate the cortical organization that accompanies human learning.

Another major requirement for the success of a GA is a valid method of evaluating each possible solution. This evaluation needs to be conducted quickly, because it must take account of many thousands of possible solutions for each generation of simulated evolution. GAs are adept at handling problems with too many variables for which to compute precise analytic solutions. The design of an engine, for example, may involve more than a hundred variables and requires satisfying dozens of constraints; GAs used by researchers at General Electric were able to come up with jet engine designs that met the constraints more precisely than conventional methods.

When using GAs you must, however, be careful what you ask for. A genetic algorithm was used to solve a block-stacking problem, and it came up with a perfect solution . . . except that it had thousands of steps. The human programmers forgot to include minimizing the number of steps in their evaluation function.

Scott Drave's Electric Sheep project is a GA that produces art. The evaluation function uses human evaluators in an open-source collaboration involving many thousands of people. The art moves through time and you can view it at electricsheep.org.

For speech recognition, the combination of genetic algorithms and hidden Markov models worked extremely well. Simulating evolution with a GA was able to substantially improve the performance of the HHMM networks. What evolution came up with was far superior to our original design, which was based on our intuition.

We then experimented with introducing a series of small variations in the overall system. For example, we would make perturbations (minor random changes) to the input. Another such change was to have adjacent Markov models "leak" into one another by causing the results of one Markov model to influence models that are "nearby." Although we did not realize it at the time, the sorts of adjustments we were experimenting with are very similar to the types of modifications that occur in biological cortical structures.

At first, such changes hurt performance (as measured by accuracy of recognition). But if we reran evolution (that is, reran the GA) with these alterations in place, it would adapt the system accordingly, optimizing it for these introduced modifications. In general, this would restore performance. If we then removed the changes we had introduced, performance would be again degraded, because the system had been evolved to compensate for the changes. The adapted system became dependent on the changes.

One type of alteration that actually helped performance (after rerunning the GA) was to introduce small random changes to the input. The reason for this is the well-known "overfitting" problem in self-organizing systems. There is a danger that such a system will overgeneralize to the specific examples contained in the training sample. By making random adjustments to the input, the more invariant patterns in the data survive, and the system thereby learns these deeper patterns. This helped only if we reran the GA with the randomization feature on.

This introduces a dilemma in our understanding of our biological cortical circuits. It had been noticed, for example, that there might indeed be a small amount of leakage from one cortical connection to another, resulting from the way that biological connections are formed: The electrochemistry of the axons and dendrites is apparently subject to the electromagnetic effects of nearby connections. Suppose we were able to run an experiment where we removed this effect in an actual brain. That would be difficult to actually carry out, but not necessarily impossible. Suppose we conducted such an experiment and found that the cortical circuits worked less effectively without this neural leakage. We might then conclude that this phenomenon was a very clever design by evolution and was critical to the cortex's achieving its level of performance. We might further point out that such a result shows that the orderly model of the flow of patterns up the conceptual hierarchy and the flow of predictions down the hierarchy was in fact much more complicated because of this intricate influence of connections on one another.

But that would not necessarily be an accurate conclusion. Consider our experience with a simulated cortex based on HHMMs, in which we implemented a modification very similar to interneuronal cross talk. If we then ran evolution with that phenomenon in place, performance would be restored (because the evolutionary process adapted to it). If we then removed the cross talk, performance would be compromised again. In the biological case, evolution (that is, biological evolution) was indeed "run" with this phenomenon in place. The detailed parameters of the system have thereby been set by biological evolution to be dependent on these factors, so that changing them will negatively affect performance unless we run evolution again. Doing so is feasible in the simulated world, where evolution only takes days or weeks, but in the biological world it would require tens of thousands of years.

So how can we tell whether a particular design feature of the biological neocortex is a vital innovation introduced by biological evolution—that is, one that is instrumental to our level of intelligence—or merely an artifact that the design of the system is now dependent on but could have evolved without? We can answer that question simply by running simulated evolution with and without these particular variations to the details of the design (for example, with and without connection cross talk). We can even do so with biological evolution if we're examining the evolution of a colony of microorganisms where generations are measured in hours, but it is not practical for complex organisms such as humans. This is another one of the many disadvantages of biology.

Getting back to our work in speech recognition, we found that if we ran evolution (that is, a GA) *separately* on the initial design of (1) the hierarchical hidden Markov models that were modeling the internal structure of phonemes and (2) the HHMMs' modeling of the structures of words and phrases, we got even better results. Both levels of the system were using HHMMs, but the GA would evolve design variations between these different levels. This approach still allowed the modeling of phenomena that occurs in between the two levels, such as the smearing of phonemes that often happens when we string certain words together (for example, "How are you all doing?" might become "How're y'all doing?").

It is likely that a similar phenomenon took place in different biological cortical regions, in that they have evolved small differences based on the types of patterns they deal with. Whereas all of these regions use the same essential neocortical algorithm, biological evolution has had enough time to fine-tune the design of each of them to be optimal for their particular patterns. However, as I discussed earlier, neuroscientists and neurologists have noticed substantial plasticity in these areas, which supports the idea of a general neocortical algorithm. If the fundamental methods in each region were radically different, then such interchangeability among cortical regions would not be possible.

The systems we created in our research using this combination of self-organizing methods were very successful. In speech recognition, they were able for the first time to handle fully continuous speech and relatively unrestricted vocabularies. We were able to achieve a high accuracy rate on a wide variety of speakers, accents, and dialects. The current state of the art as this book is being written is represented by a product called Dragon Naturally Speaking (Version 11.5) for the PC from Nuance (formerly Kurzweil Computer Products). I suggest that people try it if they are skeptical about the performance of contemporary speech recognition—accuracies are often 99 percent or higher after a few minutes of training on your voice on continuous speech and relatively unrestricted vocabularies. Dragon

Dictation is a simpler but still impressive free app for the iPhone that requires no voice training. Siri, the personal assistant on contemporary Apple iPhones, uses the same speech recognition technology with extensions to handle natural-language understanding.

The performance of these systems is a testament to the power of mathematics. With them we are essentially computing what is going on in the neocortex of a speaker—even though we have no direct access to that person's brain—as a vital step in recognizing what the person is saying and, in the case of systems like Siri, what those utterances mean. We might wonder, if we were to actually look inside the speaker's neocortex, would we see connections and weights corresponding to the hierarchical hidden Markov models computed by the software? Almost certainly we would not find a precise match; the neuronal structures would invariably differ in many details compared with the models in the computer. However, I would maintain that there must be an essential mathematical equivalence to a high degree of precision between the actual biology and our attempt to emulate it; otherwise these systems would not work as well as they do.

LISP

LISP (LISt Processor) is a computer language, originally specified by AI pioneer John McCarthy (1927–2011) in 1958. As its name suggests, LISP deals with lists. Each LISP statement is a list of elements; each element is either another list or an "atom," which is an irreducible item constituting either a number or a symbol. A list included in a list can be the list itself, hence LISP is capable of recursion. Another way that LISP statements can be recursive is if a list includes a list, and so on until the original list is specified. Because lists can include lists, LISP is also capable of hierarchical processing. A list can be a conditional such that it only "fires" if its elements are satisfied. In this way, hierarchies of such conditionals can be used to identify increasingly abstract qualities of a pattern.

LISP became the rage in the artificial intelligence community in the 1970s and early 1980s. The conceit of the LISP enthusiasts of the earlier decade was that the language mirrored the way the human brain worked— that any intelligent process could most easily and efficiently be coded in LISP. There followed a mini-boomlet in "artificial intelligence" companies that offered LISP interpreters and related LISP products, but when it became apparent in the mid-1980s that LISP itself was not a shortcut to creating intelligent processes, the investment balloon collapsed.

It turns out that the LISP enthusiasts were not entirely wrong. Essentially, each pattern recognizer in the neocortex can be regarded as a LISP statement—each one constitutes a list of elements, and each element can be another list. The neocortex is therefore indeed engaged in list processing of a symbolic nature very similar to that which takes place in a LISP program. Moreover, it processes all 300 million LISP-like "statements" simultaneously.

However, there were two important features missing from the world of LISP, one of which was learning. LISP programs had to be coded line by line by human programmers. There were attempts to automatically code LISP programs using a variety of methods, but these were not an integral part of the language's concept. The neocortex, in contrast, programs itself, filling its "statements" (that is, the lists) with meaningful and actionable information from its own experience and from its own feedback loops. This is a key principle of how the neocortex works: Each one of its pattern recognizers (that is, each LISP-like statement) is capable of filling in its own list and connecting itself both up and down to other lists. The second difference is the size parameters. One could create a variant of LISP (coded in LISP) that would allow for handling such parameters, but these are not part of the basic language.

LISP is consistent with the original philosophy of the AI field, which was to find intelligent solutions to problems and to code them directly in computer languages. The first attempt at a self-organizing method that would teach itself from experience—neural nets—was not successful because it did not provide a means to modify the topology of the system in response to learning. The hierarchical hidden Markov model effectively provided that through its pruning mechanism. Today, the HHMM together with its mathematical cousins makes up a major portion of the world of AI.

A corollary of the observation of the similarity of LISP and the list structure of the neocortex is an argument made by those who insist that the brain is too complicated for us to understand. These critics point out that the brain has trillions of connections, and since each one must be there specifically by design, they constitute the equivalent of trillions of lines of code. As we've seen, I've estimated that there are on the order of 300 million pattern processors in the neocortex—or 300 million lists where each element in the list is pointing to another list (or, at the lowest conceptual level, to a basic irreducible pattern from outside the neocortex). But 300 million is still a reasonably big number of LISP statements and indeed is larger than any human-written program in existence.

However, we need to keep in mind that these lists are not actually specified in the initial design of the nervous system. The brain creates these lists itself and connects the levels automatically from its own experiences. This is the key secret of the neocortex. The processes that accomplish this self-organization are much simpler than the 300 million statements that constitute the capacity of the neocortex. Those processes are specified in the genome. As I will demonstrate in chapter 11, the amount of unique information in the genome (after lossless compression) as applied to the brain is about 25 million bytes, which is equivalent to less than a million lines of code. The actual algorithmic complexity is even less than that, as most of the 25 million bytes of genetic information pertain to the biological needs of the neurons, and not specifically to their information-processing capability. However, even 25 million bytes of design information is a level of complexity we can handle.

Hierarchical Memory Systems

As I discussed in chapter 3, Jeff Hawkins and Dileep George in 2003 and 2004 developed a model of the neocortex incorporating hierarchical lists that was described in Hawkins and Blakeslee's 2004 book *On Intelligence*. A more up-to-date and very elegant presentation of the hierarchical temporal memory method can be found in Dileep George's 2008 doctoral dissertation.[12] Numenta has implemented it in a system called NuPIC (Numenta Platform for Intelligent Computing) and has developed pattern recognition and intelligent data-mining systems for such clients as Forbes and Power Analytics Corporation. After working at Numenta, George has started a new company called Vicarious Systems with funding from the Founder Fund (managed by Peter Thiel, the venture capitalist behind Facebook, and Sean Parker, the first president of Facebook) and from Good Ventures, led by Dustin Moskovitz, cofounder of Facebook. George reports significant progress in automatically modeling, learning, and recognizing information with a substantial number of hierarchies. He calls his system a "recursive cortical network" and plans applications for medical imaging and robotics, among other fields. The technique of hierarchical hidden Markov models is mathematically very similar to these hierarchical memory systems, especially if we allow the HHMM system to organize its own connections between pattern recognition modules. As mentioned earlier, HHMMs provide for an additional important element, which is modeling the expected distribution of the magnitude (on some continuum) of each input in computing the probability of the existence of the pattern under consideration. I have recently started a new company called Patterns, Inc., which intends to develop hierarchical self-organizing neocortical models that utilize HHMMs and related techniques for the purpose of understanding natural language. An important emphasis will be on the ability for the system to design its own hierarchies in a manner similar to a biological neocortex. Our envisioned system will continually read a wide range of material such as Wikipedia and other knowledge resources as well as listen to everything you say and watch everything you write (if you let it). The goal is for it to become a helpful friend answering your questions—*before* you even formulate them—and giving you useful information and tips as you go through your day.

The Moving Frontier of AI: Climbing the Competence Hierarchy

1. A long tiresome speech delivered by a frothy pie topping.
2. A garment worn by a child, perhaps aboard an operatic ship.
3. Wanted for a twelve-year crime spree of eating King Hrothgar's warriors; officer Beowulf has been assigned the case.
4. It can mean to develop gradually in the mind or to carry during pregnancy.
5. National Teacher Day and Kentucky Derby Day.
6. Wordsworth said they soar but never roam.
7. Four-letter word for the iron fitting on the hoof of a horse or a card-dealing box in a casino.
8. In act three of an 1846 Verdi opera, this Scourge of God is stabbed to death by his lover, Odabella.

> —Examples of *Jeopardy!* queries, all of which Watson got correct. Answers are: meringue harangue, pinafore, Grendel, gestate, May, skylark, shoe. For the eighth query, Watson replied, "What is Attila?" The host responded by saying, "Be more specific?" Watson clarified with, "What is Attila the Hun?," which is correct.

The computer's techniques for unraveling *Jeopardy!* clues sounded just like mine. That machine zeroes in on key words in a clue, then combs its memory (in Watson's case, a 15-terabyte data bank of human knowledge) for clusters of associations with these words. It rigorously checks the top hits against all the contextual information it can muster: the category name; the kind of answer being sought; the time, place, and gender hinted at in the clue; and so on. And when it feels "sure" enough, it decides to buzz. This is all an instant, intuitive process for a human *Jeopardy!* player, but I felt convinced that under the hood my brain was doing more or less the same thing.

—Ken Jennings, human *Jeopardy!*
champion who lost to Watson

I, for one, welcome our new robot overlords.

—Ken Jennings, paraphrasing *The Simpsons*,
after losing to Watson

Oh my god. [Watson] is more intelligent than the average *Jeopardy!* player in answering *Jeopardy!* questions. That's impressively intelligent.

—Sebastian Thrun, former director of
the Stanford AI Lab

Watson understands nothing. It's a bigger steamroller.

—Noam Chomsky

Artificial intelligence is all around us—we no longer have our hand on the plug. The simple act of connecting with someone via a text message, e-mail, or cell phone call uses intelligent algorithms to route the information. Almost every product we touch is originally designed in a collaboration between human and artificial intelligence and then built in automated factories. If all the AI systems decided to go on strike tomorrow, our civilization would be crippled: We couldn't get money from our bank, and indeed, our money would disappear; communication, transportation, and manufacturing would all grind to a halt. Fortunately, our intelligent machines are not yet intelligent enough to organize such a conspiracy.

What is new in AI today is the viscerally impressive nature of publicly available examples. For example, consider Google's self-driving cars (which as of this writing have gone over 200,000 miles in cities and towns), a technology that will lead to significantly fewer crashes, increased capacity of roads, alleviating the requirement of humans to perform the chore of driving, and many other benefits. Driverless cars are actually already legal to operate on public roads in Nevada with some restrictions, although widespread usage by the public throughout the world is not expected until late in this decade. Technology that intelligently watches the road and warns the driver of impending dangers is already being installed in cars. One such technology is based in part on the successful model of visual processing in the brain created by MIT's Tomaso Poggio. Called MobilEye, it was developed by Amnon Shashua, a former postdoctoral student of Poggio's. It is capable of alerting the driver to such dangers as an impending collision or a child running in front of the car and has recently been installed in cars by such manufacturers as Volvo and BMW.

I will focus in this section of the book on language technologies for several reasons. Not surprisingly, the hierarchical nature of language closely mirrors the hierarchical nature of our thinking. Spoken language was our first technology, with written language as the second. My own work in artificial intelligence, as this chapter has demonstrated, has been heavily focused on language. Finally, mastering language is a powerfully leveraged capability. Watson has already read hundreds of millions of pages on the Web and mastered the knowledge contained in these documents. Ultimately machines will be able to master all of the knowledge on the Web—which is essentially all of the knowledge of our human-machine civilization.

English mathematician Alan Turing (1912–1954) based his eponymous test on the ability of a computer to converse in natural language using text messages.[13] Turing felt that all of human intelligence was embodied and represented in language, and that no machine could pass a Turing test through simple language tricks. Although the Turing test is a game involving written language, Turing believed that the only way that a computer could pass it would be for it to actually possess the equivalent of human-level intelligence. Critics have proposed that a true test of human-level intelligence should include mastery of visual and auditory information as well.[14] Since many of my own AI projects involve teaching computers to master such sensory information as human speech, letter shapes, and musical sounds, I would be expected to advocate the inclusion of these forms of information in a true test of intelligence. Yet I agree with Turing's original insight that the text-only version of the Turing test is sufficient. Adding visual or auditory input or output to the test would not actually make it more difficult to pass.

One does not need to be an AI expert to be moved by the performance of Watson on *Jeopardy!* Although I have a reasonable understanding of the methodology used in a number of its key subsystems, that does not diminish my emotional reaction to watching it—*him?*—perform. Even a perfect understanding of how all of its component systems work—which no one actually has—would not help you to predict how Watson would actually react to a given situation. It contains hundreds of interacting subsystems, and each of these is considering millions of competing hypotheses at the same time, so predicting the outcome is impossible. Doing a thorough analysis—after the fact—of Watson's deliberations for a single three-second query would take a human centuries.

To continue my own history, in the late 1980s and 1990s we began working on natural-language understanding in limited domains. You could speak to one of our products, called Kurzweil Voice, about anything you wanted, so long as it had to do with editing documents. (For example, "Move the third paragraph on the previous page to here.") It worked pretty well in this limited but useful domain. We also created systems with medical domain knowledge so that doctors could dictate patient reports. It had enough knowledge of fields such as radiology and pathology that it could question the doctor if something in the report seemed unclear, and would guide the physician through the reporting process. These medical report-ing systems have evolved into a billion-dollar business at Nuance.

Understanding natural language, especially as an extension to auto-matic speech recognition, has now entered the mainstream. As of the writ-ing of this book, Siri, the automated personal assistant on the iPhone 4S, has created a stir in the mobile computing world. You can pretty much ask Siri to do anything that a self-respecting smartphone should be capable of doing (for example, "Where can I get some Indian food around here?" or "Text my wife that I'm on my way," or "What do people think of the new Brad Pitt movie?"), and most of the time Siri will comply. Siri will entertain a small amount of nonproductive chatter. If you ask her what the meaning of life is, she will respond with "42," which fans of *The Hitchhiker's Guide to the Galaxy* will recognize as its "answer to the ultimate question of life, the universe, and everything." Knowledge questions (including the one about the meaning of life) are answered by Wolfram Alpha, described on page 170. There is a whole world of "chatbots" who do nothing but engage in small talk. If you would like to talk to our chatbot named Ramona, go to our Web site KurzweilAI.net and click on "Chat with Ramona."

Some people have complained to me about Siri's failure to answer cer-tain requests, but I often recall that these are the same people who persis-tently complain about human service providers also. I sometimes suggest that we try it together, and often it works better than they expect. The com-plaints remind me of the story of the dog who plays chess. To an incredu-lous questioner, the dog's owner replies, "Yeah, it's true, he does play chess, but his endgame is weak." Effective competitors are now emerging, such as Google Voice Search.

That the general public is now having conversations in natural spoken language with their handheld computers marks a new era. It is typical that people dismiss the significance of a first-generation technology because of its limitations. A few years later, when the technology does work well, peo-ple still dismiss its importance because, well, it's no longer new. That being said, Siri works impressively for a first-generation product, and it is clear that this category of product is only going to get better.

Siri uses the HMM-based speech recognition technologies from Nuance. The natural-language extensions were first developed by the DARPA-funded "CALO" project.[15] Siri has been enhanced with Nuance's own natural-language technologies, and Nuance offers a very similar technology called Dragon Go![16]

The methods used for understanding natural language are very similar to hierarchical hidden Markov models, and indeed HHMM itself is commonly used. Whereas some of these systems are not specifically labeled as using HMM or HHMM, the mathematics is virtually identical. They all involve hierarchies of linear sequences where each element has a weight, connections that are self-adapting, and an overall system that self-organizes based on learning data. Usually the learning continues during actual use of the system. This approach matches the hierarchical structure of natural language—it is just a natural extension up the conceptual ladder from parts of speech to words to phrases to semantic structures. It would make sense to run a genetic algorithm on the parameters that control the precise learning algorithm of this class of hierarchical learning systems and determine the optimal algorithmic details.

Over the past decade there has been a shift in the way that these hierarchical structures are created. In 1984 Douglas Lenat (born in 1950) started the ambitious Cyc (for enCYClopedic) project, which aimed to create rules that would codify everyday "commonsense" knowledge. The rules were organized in a huge hierarchy, and each rule involved—again—a linear sequence of states. For example, one Cyc rule might state that a dog has a face. Cyc can then link to general rules about the structure of faces: that a face has two eyes, a nose, and a mouth, and so on. We don't need to have one set of rules for a dog's face and then another for a cat's face, though we may of course want to put in additional rules for ways in which dogs' faces differ from cats' faces. The system also includes an inference engine: If we have rules that state that a cocker spaniel is a dog, that dogs are animals, and that animals eat food, and if we were to ask the inference engine whether cocker spaniels eat, the system would respond that yes, cocker spaniels eat food. Over the next twenty years, and with thousands of person-years of effort, over a million such rules were written and tested. Interestingly, the language for writing Cyc rules—called CycL—is almost identical to LISP.

Meanwhile, an opposing school of thought believed that the best approach to natural-language understanding, and to creating intelligent systems in general, was through automated learning from exposure to a very large number of instances of the phenomena the system was trying to master. A powerful example of such a system is Google Translate, which can translate to and from fifty languages. That's 2,500 different translation directions, although for most language pairs, rather than translate language 1 directly into language 2, it will translate language 1 into English and then English into language 2. That reduces the number of translators Google needed to build to ninety-eight (plus a limited number of non-English pairs for which there is direct translation). The Google translators do not use grammatical rules; rather, they create vast databases for each language pair of common translations based on large "Rosetta stone" corpora of translated documents between two languages. For the six languages that constitute the official languages of the United Nations, Google has used United Nations documents, as they are published in all six languages. For less common languages, other sources have been used.

The results are often impressive. DARPA runs annual competitions for the best automated language translation systems for different language pairs, and Google Translate often wins for certain pairs, outperforming systems created directly by human linguists.

Over the past decade two major insights have deeply influenced the natural-language-understanding field. The first has to do with hierarchies. Although the Google approach started with association of flat word sequences from one language to another, the inherent hierarchical nature of language has inevitably crept into its operation. Systems that methodically incorporate hierarchical learning (such as hierarchical hidden Markov models) provided significantly better performance. However, such systems are not quite as automatic to build. Just as humans need to learn approximately one conceptual hierarchy at a time, the same is true for computerized systems, so the learning process needs to be carefully managed.

The other insight is that hand-built rules work well for a core of common basic knowledge. For translations of short passages, this approach often provides more accurate results. For example, DARPA has rated rule-based Chinese-to-English translators higher than Google Translate for short passages. For what is called the tail of a language, which refers to the millions of infrequent phrases and concepts used in it, the accuracy of rule-based systems approaches an unacceptably low asymptote. If we plot natural-language-understanding accuracy against the amount of training data analyzed, rule-based systems have higher performance initially but level off at fairly low accuracies of about 70 percent. In sharp contrast, statistical systems can reach the high 90s in accuracy but require a great deal of data to achieve that.

Often we need a combination of at least moderate performance on a small amount of training data and then the opportunity to achieve high accuracies with a more significant quantity. Achieving moderate performance quickly enables us to put a system in the field and then to automatically collect training data as people actually use it. In this way, a great deal of learning can occur at the same time that the system is being used, and its accuracy will improve. The statistical learning needs to be fully hierarchical to reflect the nature of language, which also reflects how the human brain works.

This is also how Siri and Dragon Go! work—using rules for the most common and reliable phenomena and then learning the "tail" of the language in the hands of real users. When the Cyc team realized that they had reached a ceiling of performance based on hand-coded rules, they too adopted this approach. Hand-coded rules provide two essential functions.

They offer adequate initial accuracy, so that a trial system can be placed into widespread usage, where it will improve automatically. Secondly, they provide a solid basis for the lower levels of the conceptual hierarchy so that the automated learning can begin to learn higher conceptual levels.

As mentioned above, Watson represents a particularly impressive example of the approach of combining hand-coded rules with hierarchical statistical learning. IBM combined a number of leading natural-language programs to create a system that could play the natural-language game of *Jeopardy!* On February 14–16, 2011, Watson competed with the two leading human players: Brad Rutter, who had won more money than anyone else on the quiz show, and Ken Jennings, who had previously held the *Jeopardy!* championship for the record time of seventy-five days.

By way of context, I had predicted in my first book, *The Age of Intelligent Machines,* written in the mid-1980s, that a computer would take the world chess championship by 1998. I also predicted that when that

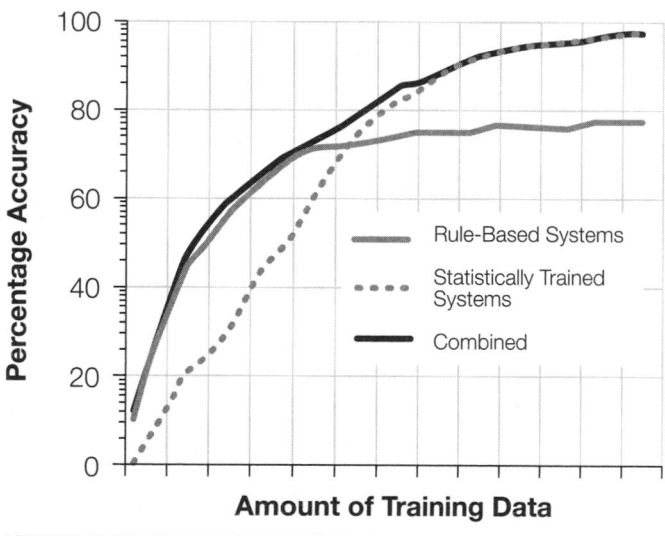

The accuracy of natural-language-understanding systems as a function of the amount of training data. The best approach is to combine rules for the "core" of the language and a data-based approach for the "tail" of the language.

happened, we would either downgrade our opinion of human intelligence, upgrade our opinion of machine intelligence, or downplay the importance of chess, and that if history was a guide, we would minimize chess. Both of these things happened in 1997. When IBM's chess supercomputer Deep Blue defeated the reigning human world chess champion, Garry Kasparov, we were immediately treated to arguments that it was to be expected that a computer would win at chess because computers are logic machines, and chess, after all, is a game of logic. Thus Deep Blue's victory was judged to be neither surprising nor significant. Many of its critics went on to argue that computers would never master the subtleties of human language, including metaphors, similes, puns, double entendres, and humor.

That is at least one reason why Watson represents such a significant milestone: *Jeopardy!* is precisely such a sophisticated and challenging language task. Typical *Jeopardy!* queries includes many of these vagaries of human language. What is perhaps not evident to many observers is that Watson not only had to master the language in the unexpected and convoluted queries, but for the most part its knowledge was not hand-coded. It obtained that knowledge by actually reading 200 million pages of natural-language documents, including all of Wikipedia and other encyclopedias, comprising 4 trillion bytes of language-based knowledge. As readers of this book are well aware, Wikipedia is not written in LISP or CycL, but rather in natural sentences that have all of the ambiguities and intricacies inherent in language. Watson needed to consider all 4 trillion characters in its reference material when responding to a question. (I realize that *Jeopardy!* queries are answers in search of a question, but this is a technicality— they ultimately are really questions.) If Watson can understand and respond to questions based on 200 million pages—in three seconds!— there is nothing to stop similar systems from reading the other billions of documents on the Web. Indeed, that effort is now under way.

When we were developing character and speech recognition systems and early natural-language-understanding systems in the 1970s through 1990s, we used a methodology of incorporating an "expert manager." We would develop multiple systems to do the same thing but would incorporate somewhat different approaches in each one. Some of the differences were subtle, such as variations in the parameters controlling the mathematics of the learning algorithm. Some variations were fundamental, such as including rule-based systems instead of hierarchical statistical learning systems. The expert manager was itself a software program that was programmed to learn the strengths and weaknesses of these different systems by examining their performance in real-world situations. It was based on the notion that these strengths were orthogonal; that is, one system would tend to be strong where another was weak. Indeed, the overall performance of the combined systems with the trained expert manager in charge was far better than any of the individual systems.

Watson works the same way. Using an architecture called UIMA (Unstructured Information Management Architecture), Watson deploys literally hundreds of different systems—many of the individual language components in Watson are the same ones that are used in publicly available natural-language-understanding systems—all of which are attempting to either directly come up with a response to the *Jeopardy!* query or else at least provide some disambiguation of the query. UIMA is basically acting as the expert manager to intelligently combine the results of the independent systems. UIMA goes substantially beyond earlier systems, such as the one we developed in the predecessor company to Nuance, in that its individual systems can contribute to a result without necessarily coming up with a final answer. It is sufficient if a subsystem helps narrow down the solution. UIMA is also able to compute how much confidence it has in the final answer. The human brain does this also—we are probably very confident of our response when asked for our mother's first name, but we are less so in coming up with the name of someone we met casually a year ago.

Thus rather than come up with a single elegant approach to understanding the language problem inherent in *Jeopardy!* the IBM scientists combined all of the state-of-the-art language-understanding modules they could get their hands on. Some use hierarchical hidden Markov models;

some use mathematical variants of HHMM; others use rule-based approaches to code directly a core set of reliable rules. UIMA evaluates the performance of each system in actual use and combines them in an optimal way. There is some misunderstanding in the public discussions of Watson in that the IBM scientists who created it often focus on UIMA, which is the expert manager they created. This leads to comments by some observers that Watson has no real understanding of language because it is difficult to identify where this understanding resides. Although the UIMA framework also learns from its own experience, Watson's "understanding" of language cannot be found in UIMA alone but rather is distributed across all of its many components, including the self-organizing language modules that use methods similar to HHMM.

A separate part of Watson's technology uses UIMA's confidence estimate in its answers to determine how to place *Jeopardy!* bets. While the Watson system is specifically optimized to play this particular game, its core language- and knowledge-searching technology can easily be adapted to other broad tasks. One might think that less commonly shared professional knowledge, such as that in the medical field, would be more difficult to master than the general-purpose "common" knowledge that is required to play *Jeopardy!* Actually, the opposite is the case: Professional knowledge tends to be more highly organized, structured, and less ambiguous than its commonsense counterpart, so it is highly amenable to accurate natural-language understanding using these techniques. As mentioned, IBM is currently working with Nuance to adapt the Watson technology to medicine.

The conversation that takes place when Watson is playing *Jeopardy!* is a brief one: A question is posed, and Watson comes up with an answer. (Again, technically, it comes up with a question to respond to an answer.) It does not engage in a conversation that would require tracking all of the earlier statements of all participants. (Siri actually does do this to a limited extent: If you ask it to send a message to your wife, it will ask you to identify her, but it will remember who she is for subsequent requests.) Tracking all of the information in a conversation—a task that would clearly be required to pass the Turing test—is a significant additional requirement but not fundamentally more difficult than what Watson is doing already. After all, Watson has read hundreds of millions of pages of material, which obviously includes many stories, so it is capable of tracking through complicated sequential events. It should therefore be able to follow its own conversations and take that into consideration in its subsequent replies.

Another limitation of the *Jeopardy!* game is that the answers are generally brief: It does not, for example, pose questions of the sort that ask contestants to name the five primary themes of *A Tale of Two Cities*. To the extent that it can find documents that do discuss the themes of this novel, a suitably modified version of Watson should be able to respond to this. Coming up with such themes on its own from just reading the book, and not essentially copying the thoughts (even without the words) of other thinkers, is another matter. Doing so would constitute a higher-level task than Watson is capable of today—it is what I call a Turing test–level task. (That being said, I will point out that most humans do not come up with their own original thoughts either but copy the ideas of their peers and opinion leaders.) At any rate, this is 2012, not 2029, so I would not expect Turing test–level intelligence yet. On yet another hand, I would point out that evaluating the answers to questions such as finding key ideas in a novel is itself not a straightforward task. If someone is asked who signed the Declaration of Independence, one can determine whether or not her response is true or false. The validity of answers to higher-level questions such as describing the themes of a creative work is far less easily established.

It is noteworthy that although Watson's language skills are actually somewhat below that of an educated human, it was able to defeat the best two *Jeopardy!* players in the world. It could accomplish this because it is able to combine its language ability and knowledge understanding with the perfect recall and highly accurate memories that machines possess. That is why we have already largely assigned our personal, social, and historical memories to them.

Although I'm not prepared to move up my prediction of a computer passing the Turing test by 2029, the progress that has been achieved in systems like Watson should give anyone substantial confidence that the advent of Turing-level AI is close at hand. If one were to create a version of Watson that was optimized for the Turing test, it would probably come pretty close.

American philosopher John Searle (born in 1932) argued recently that Watson is not capable of thinking. Citing his "Chinese room" thought experiment (which I will discuss further in chapter 11), he states that Watson is only manipulating symbols and does not understand the meaning of those symbols. Actually, Searle is not describing Watson accurately, since its understanding of language is based on hierarchical statistical processes—not the manipulation of symbols. The only way that Searle's characterization would be accurate is if we considered every step in Watson's self-organizing processes to be "the manipulation of symbols." But if that were the case, then the human brain would not be judged capable of thinking either.

It is amusing and ironic when observers criticize Watson for *just* doing statistical analysis of language as opposed to possessing the "true" understanding of language that humans have. Hierarchical statistical analysis is exactly what the human brain is doing when it is resolving multiple hypotheses based on statistical inference (and indeed at every level of the neocortical hierarchy). Both Watson and the human brain learn and respond based on a similar approach to hierarchical understanding. In many respects Watson's knowledge is far more extensive than a human's; no human can claim to have mastered all of Wikipedia, which is only part of Watson's knowledge base. Conversely, a human can today master more conceptual levels than Watson, but that is certainly not a permanent gap.

One important system that demonstrates the strength of computing applied to organized knowledge is Wolfram Alpha, an answer engine (as opposed to a search engine) developed by British mathematician and scientist Dr. Wolfram (born 1959) and his colleagues at Wolfram Research.

For example, if you ask Wolfram Alpha (at WolframAlpha.com), "How many primes are there under a million?" it will respond with "78,498." It did not look up the answer, it computed it, and following the answer it provides the equations it used. If you attempted to get that answer using a conventional search engine, it would direct you to links where you could find the algorithms required. You would then have to plug those formulas into a system such as Mathematica, also developed by Dr. Wolfram, but this would obviously require a lot more work (and understanding) than simply asking Alpha.

Indeed, Alpha consists of 15 million lines of Mathematica code. What Alpha is doing is literally computing the answer from approximately 10 trillion bytes of data that have been carefully curated by the Wolfram Research staff. You can ask a wide range of factual questions, such as "What country has the highest GDP per person?" (Answer: Monaco, with $212,000 per person in U.S. dollars), or "How old is Stephen Wolfram?" (Answer: 52 years, 9 months, 2 days as of the day I am writing this). As mentioned, Alpha is used as part of Apple's Siri; if you ask Siri a factual question, it is handed off to Alpha to handle. Alpha also handles some of the searches posed to Microsoft's Bing search engine.

In a recent blog post, Dr. Wolfram reported that Alpha is now providing successful responses 90 percent of the time.[17] He also reports an exponential decrease in the failure rate, with a half-life of around eighteen months. It is an impressive system, and uses handcrafted methods and hand-checked data. It is a testament to why we created computers in the first place. As we discover and compile scientific and mathematical methods, computers are far better than unaided human intelligence in implementing them. Most of the known scientific methods have been encoded in Alpha, along with continually updated data on topics ranging from economics to physics. In a private conversation I had with Dr. Wolfram, he estimated that self-organizing methods such as those used in Watson typically achieve about an 80 percent accuracy when they are working well. Alpha, he pointed out, is achieving about a 90 percent accuracy. Of course, there is self-selection in both of these accuracy numbers in that users (such as myself) have learned what kinds of questions Alpha is good at, and a similar factor applies to the self-organizing methods. Eighty percent appears to be a reasonable estimate of how accurate Watson is on *Jeopardy!* queries, but this was sufficient to defeat the best humans.

It is my view that self-organizing methods such as I articulated in the pattern recognition theory of mind are needed to understand the elaborate and often ambiguous hierarchies we encounter in real-world phenomena, including human language. An ideal combination for a robustly intelligent system would be to combine hierarchical intelligence based on the PRTM (which I contend is how the human brain works) with precise codification of scientific knowledge and data. That essentially describes a human with a computer. We will enhance both poles of intelligence in the years ahead. With regard to our biological intelligence, although our neocortex has significant plasticity, its basic architecture is limited by its physical constraints. Putting additional neocortex into our foreheads was an important evolutionary innovation, but we cannot now easily expand the size of our frontal lobes by a factor of a thousand, or even by 10 percent. That is, we cannot do so biologically, but that is exactly what we will do technologically.

A Strategy for Creating a Mind

> There are billions of neurons in our brains, but what are neurons? Just cells. The brain has no knowledge until connections are made between neurons. All that we know, all that we are, comes from the way our neurons are connected.
>
> —Tim Berners-Lee

Let's use the observations I have discussed above to begin building a brain. We will start by building a pattern recognizer that meets the necessary attributes. Next we'll make as many copies of the recognizer as we have memory and computational resources to support. Each recognizer computes the probability that its pattern has been recognized. In doing so, it takes into consideration the observed magnitude of each input (in some appropriate continuum) and matches these against the learned size and size variability parameters associated with each input. The recognizer triggers its simulated axon if that computed probability exceeds a threshold. This threshold and the parameters that control the computation of the pattern's probability are among the parameters we will optimize with a genetic algorithm. Because it is not a requirement that every input be active for a pattern to be recognized, this provides for autoassociative recognition (that is, recognizing a pattern based on only part of the pattern being present). We also allow for inhibitory signals (signals that indicate that the pattern is less likely).

Recognition of the pattern sends an active signal up the simulated axon of this pattern recognizer. This axon is in turn connected to one or more pattern recognizers at the next higher conceptual level. All of the pattern recognizers connected at the next higher conceptual level are accepting this pattern as one of its inputs. Each pattern recognizer also sends signals down to pattern recognizers at lower conceptual levels whenever most of a pattern has been recognized, indicating that the rest of the pattern is "expected." Each pattern recognizer has one or more of these expected signal input channels. When an expected signal is received in this way, the threshold for recognition of this pattern recognizer is lowered (made easier).

The pattern recognizers are responsible for "wiring" themselves to other pattern recognizers up and down the conceptual hierarchy. Note that all the "wires" in a software implementation operate via virtual links (which, like Web links, are basically memory pointers) and not actual wires. This system is actually much more flexible than that in the biological brain. In a human brain, new patterns have to be assigned to an actual physical pattern recognizer, and new connections have to be made with an actual axon-to-dendrite link. Usually this means taking an existing physical connection that is approximately what is needed and then growing the necessary axon and dendrite extensions to complete the full connection.

Another technique used in biological mammalian brains is to start with a large number of possible connections and then prune the neural connections that are not used. If a biological neocortex reassigns cortical pattern recognizers that have already learned older patterns in order to learn more recent material, then the connections need to be physically reconfigured. Again, these tasks are much simpler in a software implementation. We simply assign new memory locations to a new pattern recognizer and use memory links for the connections. If the digital neocortex wishes to reassign cortical memory resources from one set of patterns to another, it simply returns the old pattern recognizers to memory and then makes the new assignment. This sort of "garbage collection" and reassignment of memory is a standard feature of the architecture of many software systems. In our digital brain we would also back up old memories before discarding them from the active neocortex, a precaution we can't take in our biological brains.

There are a variety of mathematical techniques that can be employed to implement this approach to self-organizing hierarchical pattern recognition. The method I would use is hierarchical hidden Markov models, for several reasons. From my personal perspective, I have several decades of familiarity with this method, having used it in the earliest speech recognition and natural-language systems starting in the 1980s. From the perspective of the overall field, there is greater experience with hidden Markov models than with any other approach for pattern recognition tasks. They are also extensively used in natural-language understanding. Many NLU systems use techniques that are at least mathematically similar to HHMM.

Note that not all hidden Markov model systems are fully hierarchical. Some allow for just a few levels of hierarchy—for example, going from acoustic states to phonemes to words. To build a brain, we will want to enable our system to create as many new levels of hierarchy as needed.

Also, most hidden Markov model systems are not fully self-organizing. Some have fixed connections, although these systems do effectively prune many of their starting connections by allowing them to evolve zero connection weights. Our systems from the 1980s and 1990s automatically pruned connections with connection weights below a certain level and also allowed for making new connections to better model the training data and to learn on the fly. A key requirement, I believe, is to allow for the system to flexibly create its own topologies based on the patterns it is exposed to while learning. We can use the mathematical technique of linear programming to optimally assign connections to new pattern recognizers.

Our digital brain will also accommodate substantial redundancy of each pattern, especially ones that occur frequently. This allows for robust recognition of common patterns and is also one of the key methods to achieving invariant recognition of different forms of a pattern. We will, however, need rules for how much redundancy to permit, as we don't want to use up excessive amounts of memory on very common low-level patterns.

The rules regarding redundancy, recognition thresholds, and the effect on the threshold of a "this pattern is expected" indication are a few examples of key overall parameters that affect the performance of this type of self-organizing system. I would initially set these parameters based on my intuition, but we would then optimize them using a genetic algorithm.

A very important consideration is the education of a brain, whether a biological or a software one. As I discussed earlier, a hierarchical pattern recognition system (digital or biological) will only learn about two—preferably one—hierarchical levels at a time. To bootstrap the system I would start with previously trained hierarchical networks that have already learned their lessons in recognizing human speech, printed characters, and natural-language structures. Such a system would be capable of reading natural-language documents but would only be able to master approximately one conceptual level at a time. Previously learned levels would provide a relatively stable basis to learn the next level. The system can read the same documents over and over, gaining new conceptual levels with each subsequent reading, similar to the way people reread and achieve a deeper understanding of texts. Billions of pages of material are available on the Web. Wikipedia itself has about four million articles in the English version.

I would also provide a critical thinking module, which would perform a continual background scan of all of the existing patterns, reviewing their compatibility with the other patterns (ideas) in this software neocortex. We have no such facility in our biological brains, which is why people can hold completely inconsistent thoughts with equanimity. Upon identifying an inconsistent idea, the digital module would begin a search for a resolution, including its own cortical structures as well as all of the vast literature available to it. A resolution might simply mean determining that one of the inconsistent ideas is simply incorrect (if contraindicated by a preponderance of conflicting data). More constructively, it would find an idea at a higher conceptual level that resolves the apparent contradiction by providing a perspective that explains each idea. The system would add this resolution as a new pattern and link to the ideas that initially triggered the search for the resolution. This critical thinking module would run as a continual background task. It would be very beneficial if human brains did the same thing.

I would also provide a module that identifies open questions in every discipline. As another continual background task, it would search for solutions to them in other disparate areas of knowledge. As I noted, the knowledge in the neocortex consists of deeply nested patterns of patterns and is therefore entirely metaphorical. We can use one pattern to provide a solution or insight in an apparently disconnected field.

As an example, recall the metaphor I used in chapter 4 relating the random movements of molecules in a gas to the random movements of evolutionary change. Molecules in a gas move randomly with no apparent sense of direction. Despite this, virtually every molecule in a gas in a beaker, given sufficient time, will leave the beaker. I noted that this provides a perspective on an important question concerning the evolution of intelligence. Like molecules in a gas, evolutionary changes also move every which way with no apparent direction. Yet we nonetheless see a movement toward greater complexity and greater intelligence, indeed to evolution's supreme achievement of evolving a neocortex capable of hierarchical thinking. So we are able to gain an insight into how an apparently purposeless and directionless process can achieve an apparently purposeful result in one field (biological evolution) by looking at another field (thermodynamics).

I mentioned earlier how Charles Lyell's insight that minute changes to rock formations by streaming water could carve great valleys over time inspired Charles Darwin to make a similar observation about continual minute changes to the characteristics of organisms within a species. This metaphor search would be another continual background process.

We should provide a means of stepping through multiple lists simultaneously to provide the equivalent of structured thought. A list might be the statement of the constraints that a solution to a problem must satisfy. Each step can generate a recursive search through the existing hierarchy of ideas or a search through available literature. The human brain appears to be able to handle only four simultaneous lists at a time (without the aid of tools such as computers), but there is no reason for an artificial neocortex to have such a limitation.

We will also want to enhance our artificial brains with the kind of intelligence that computers have always excelled in, which is the ability to master vast databases accurately and implement known algorithms quickly and efficiently. Wolfram Alpha uniquely combines a great many known scientific methods and applies them to carefully collected data. This type of system is also going to continue to improve given Dr. Wolfram's observation of an exponential decline in error rates.

Finally, our new brain needs a purpose. A purpose is expressed as a series of goals. In the case of our biological brains, our goals are established by the pleasure and fear centers that we have inherited from the old brain. These primitive drives were initially set by biological evolution to foster the survival of species, but the neocortex has enabled us to sublimate them. Watson's goal was to respond to *Jeopardy!* queries. Another simply stated goal could be to pass the Turing test. To do so, a digital brain would need a human narrative of its own fictional story so that it can pretend to be a biological human. It would also have to dumb itself down considerably, for any system that displayed the knowledge of, say, Watson would be quickly unmasked as nonbiological.

More interestingly, we could give our new brain a more ambitious goal, such as contributing to a better world. A goal along these lines, of course, raises a lot of questions: Better for whom? Better in what way? For biological humans? For all conscious beings? If that is the case, who or what is conscious?

As nonbiological brains become as capable as biological ones of effecting changes in the world—indeed, ultimately far more capable than unenhanced biological ones—we will need to consider their moral education. A good place to start would be with one old idea from our religious traditions: the golden rule.

CHAPTER 8

THE MIND AS COMPUTER

Shaped a little like a loaf of French country bread, our brain is a crowded chemistry lab, bustling with nonstop neural conversations. Imagine the brain, that shiny mound of being, that mouse-gray parliament of cells, that dream factory, that petit tyrant inside a ball of bone, that huddle of neurons calling all the plays, that little everywhere, that fickle pleasuredome, that wrinkled wardrobe of selves stuffed into the skull like too many clothes into a gym bag.

—Diane Ackerman

Brains exist because the distribution of resources necessary for survival and the hazards that threaten survival vary in space and time.

—John M. Allman

The modern geography of the brain has a deliciously antiquated feel to it—rather like a medieval map with the known world encircled by terra incognita where monsters roam.

—David Bainbridge

In mathematics you don't understand things. You just get used to them.

—John von Neumann

Ever since the emergence of the computer in the middle of the twentieth century, there has been ongoing debate not only about the ultimate extent of its abilities but about whether the human brain itself could be considered a form of computer. As far as the latter question was concerned, the consensus has veered from viewing these two kinds of information-processing entities as being essentially the same to their being fundamentally different. So *is* the brain a computer?

When computers first became a popular topic in the 1940s, they were immediately regarded as thinking machines. The ENIAC, which was announced in 1946, was described in the press as a "giant brain." As computers became commercially available in the following decade, ads routinely referred to them as brains capable of feats that ordinary biological brains could not match.

A 1957 ad showing the popular conception of a computer as a giant brain.

Computer programs quickly enabled the machines to live up to this billing. The "general problem solver," created in 1959 by Herbert A. Simon, J. C. Shaw, and Allen Newell at Carnegie Mellon University, was able to devise a proof to a theorem that mathematicians Bertrand Russell (1872–1970) and Alfred North Whitehead (1861–1947) had been unable to solve in their famous 1913 work *Principia Mathematica*. What became apparent in the decades that followed was that computers could readily significantly exceed unassisted human capability in such intellectual exercises as solving mathematical problems, diagnosing disease, and playing chess but had difficulty with controlling a robot tying shoelaces or with understanding the commonsense language that a five-year-old child could comprehend. Computers are only now starting to master these sorts of skills. Ironically, the evolution of computer intelligence has proceeded in the opposite direction of human maturation.

The issue of whether or not the computer and the human brain are at some level equivalent remains controversial today. In the introduction I mentioned that there were millions of links for quotations on the complexity of the human brain. Similarly, a Google inquiry for "Quotations: the brain is not a computer" also returns millions of links. In my view, statements along these lines are akin to saying, "Applesauce is not an apple." Technically that statement is true, but you can make applesauce from an apple. Perhaps more to the point, it is like saying, "Computers are not word processors." It is true that a computer and a word processor exist at different conceptual levels, but a computer can become a word processor if it is running word processing software and not otherwise. Similarly, a computer can become a brain if it is running brain software. That is what researchers including myself are attempting to do.

The question, then, is whether or not we can find an algorithm that would turn a computer into an entity that is equivalent to a human brain. A computer, after all, can run any algorithm that we might define because of its innate universality (subject only to its capacity). The human brain, on the other hand, is running a specific set of algorithms. Its methods are clever in that it allows for significant plasticity and the restructuring of its own connections based on its experience, but these functions can be emulated in software.

The universality of computation (the concept that a general-purpose computer can implement any algorithm)—and the power of this idea—emerged at the same time as the first actual machines. There are four key concepts that underlie the universality and feasibility of computation and its applicability to our thinking. They are worth reviewing here, because the brain itself makes use of them. The first is the ability to communicate, remember, and compute information reliably. Around 1940, if you used the word "computer," people assumed you were talking about an analog computer, in which numbers were represented by different levels of voltage, and specialized components could perform arithmetic functions such as addition and multiplication. A big limitation of analog computers, however, was that they were plagued by accuracy issues. Numbers could only be represented with an accuracy of about one part in a hundred, and as voltage levels representing them were processed by increasing numbers of arithmetic operators, errors would accumulate. If you wanted to perform more than a handful of computations, the results would become so inaccurate as to be meaningless.

Anyone who can remember the days of recording music with analog tape machines will recall this effect. There was noticeable degradation on the first copy, as it was a little noisier than the original. (Remember that "noise" represents random inaccuracies.) A copy of the copy was noisier still, and by the tenth generation the copy was almost entirely noise. It was assumed that the same problem would plague the emerging world of digital computers. We can understand such concerns if we consider the communication of digital information through a channel. No channel is perfect and each one will have some inherent error rate. Suppose we have a channel that has a .9 probability of correctly transmitting each bit. If I send a message that is one bit long, the probability of accurately transmitting it through that channel will be .9. Suppose I send two bits? Now the accuracy is $.9^2 = .81$. How about if I send one byte (eight bits)? I have less than an even chance (.43 to be exact) of sending it correctly. The probability of accurately sending five bytes is about 1 percent.

An obvious solution to circumvent this problem is to make the channel more accurate. Suppose the channel makes only one error in a million bits. If I send a file consisting of a half million bytes (about the size of a modest program or database), the probability of correctly transmitting it is less than 2 percent, despite the very high inherent accuracy of the channel. Given that a single-bit error can completely invalidate a computer program and other forms of digital data, that is not a satisfactory situation. Regardless of the accuracy of the channel, since the likelihood of an error in a transmission grows rapidly with the size of the message, this would seem to be an intractable barrier.

Analog computers approached this problem through graceful degradation (meaning that users only presented problems in which they could tolerate small errors); however, if users of analog computers limited themselves to a constrained set of calculations, the computers did prove somewhat useful. Digital computers, on the other hand, require continual communication, not just from one computer to another, but within the computer itself. There is communication from its memory to and from the central processing unit. Within the central processing unit, there is communication from one register to another and back and forth to the arithmetic unit, and so forth. Even within the arithmetic unit, there is communication from one bit register to another. Communication is pervasive at every level. If we consider that error rates escalate rapidly with increased communication and that a single-bit error can destroy the integrity of a process, digital computation was doomed—or so it seemed at the time.

Remarkably, that was the common view until American mathematician Claude Shannon (1916–2001) came along and demonstrated how we can create arbitrarily accurate communication using even the most unreliable communication channels. What Shannon stated in his landmark paper "A Mathematical Theory of Communication," published in the *Bell System Technical Journal* in July and October 1948, and in particular in his noisy channel-coding theorem, was that if you have available a channel with any error rate (except for exactly 50 percent per bit, which would mean that the channel was just transmitting pure noise), you are able to transmit a message in which the error rate is as accurate as you desire. In other words, the error rate of the transmission can be one bit out of *n* bits, where *n* can be as large as you define. So, for example, in the extreme, if you have a channel that correctly transmits bits of information only 51 percent of the time (that is, it transmits the correct bit just slightly more often than the wrong bit), you can nonetheless transmit messages such that only one bit out of a million is incorrect, or one bit out of a trillion or a trillion trillion.

How is this possible? The answer is through redundancy. That may seem obvious now, but it was not at the time. As a simple example, if I transmit each bit three times and take the majority vote, I will have substantially increased the reliability of the result. If that is not good enough, simply increase the redundancy until you get the reliability you need. Simply repeating information is the easiest way to achieve arbitrarily high accuracy rates from low-accuracy channels, but it is not the most efficient approach. Shannon's paper, which established the field of information theory, presented optimal methods of error detection and correction codes that can achieve *any* target accuracy through *any* nonrandom channel.

Older readers will recall telephone modems, which transmitted information through noisy analog phone lines. These lines featured audibly obvious hisses and pops and many other forms of distortion, but nonetheless were able to transmit digital data with very high accuracy rates, thanks to Shannon's noisy channel theorem. The same issue and the same solution exist for digital memory. Ever wonder how CDs, DVDs, and program disks continue to provide reliable results even after the disk has been dropped on the floor and scratched? Again, we can thank Shannon.

Computation consists of three elements: communication—which, as I mentioned, is pervasive both within and between computers—memory, and logic gates (which perform the arithmetic and logical functions). The accuracy of logic gates can also be made arbitrarily high by similarly using error detection and correction codes. It is due to Shannon's theorem and theory that we can handle arbitrarily large and complex digital data and algorithms without the processes being disturbed or destroyed by errors. It is important to point out that the brain uses Shannon's principle as well, although the evolution of the human brain clearly predates Shannon's own! Most of the patterns or ideas (and an idea is also a pattern), as we have seen, are stored in the brain with a substantial amount of redundancy. A primary reason for the redundancy in the brain is the inherent unreliability of neural circuits.

The second important idea on which the information age relies is the one I mentioned earlier: the universality of computation. In 1936 Alan Turing described his "Turing machine," which was not an actual machine but another thought experiment. His theoretical computer consists of an infinitely long memory tape with a 1 or a 0 in each square. Input to the machine is presented on this tape, which the machine can read one square at a time. The machine also contains a table of rules—essentially a stored program—that consist of numbered states. Each rule specifies one action if the square currently being read is a 0, and a different action if the current square is a 1. Possible actions include writing a 0 or 1 on the tape, moving the tape one square to the right or left, or halting. Each state will then specify the number of the next state that the machine should be in.

The input to the Turing machine is presented on the tape. The program runs, and when the machine halts, it has completed its algorithm, and the output of the process is left on the tape. Note that even though the tape is theoretically infinite in length, any actual program that does not get into an infinite loop will use only a finite portion of the tape, so if we limit ourselves to a finite tape, the machine will still solve a useful set of problems.

If the Turing machine sounds simple, it is because that was its inventor's objective. Turing wanted his machine to be as simple as possible (but

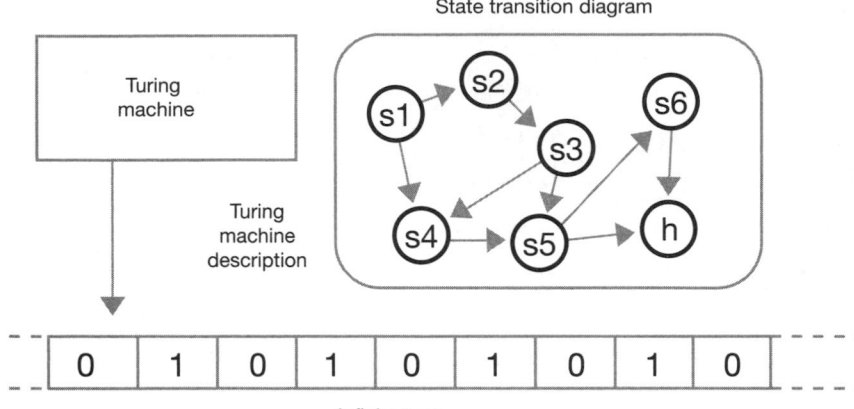

State transition diagram

A block diagram of a Turing machine with a head that reads and writes the tape and an internal program consisting of state transitions.

no simpler, to paraphrase Einstein). Turing and Alonzo Church (1903–1995), his former professor, went on to develop the Church-Turing thesis, which states that if a problem that can be presented to a Turing machine is not solvable by it, it is also not solvable by *any* machine, following natural law. Even though the Turing machine has only a handful of commands and processes only one bit at a time, it can compute anything that any computer can compute. Another way to say this is that any machine that is "Turing complete" (that is, that has equivalent capabilities to a Turing machine) can compute any algorithm (any procedure that we can define).

"Strong" interpretations of the Church-Turing thesis propose an essential equivalence between what a human can think or know and what is computable by a machine. The basic idea is that the human brain is likewise subject to natural law, and thus its information-processing ability cannot exceed that of a machine (and therefore of a Turing machine).

We can properly credit Turing with establishing the theoretical foundation of computation with his 1936 paper, but it is important to note that he was deeply influenced by a lecture that Hungarian American mathematician John von Neumann (1903–1957) gave in Cambridge in 1935 on his stored program concept, a concept enshrined in the Turing machine.[1] In turn, von Neumann was influenced by Turing's 1936 paper, which elegantly laid out the principles of computation, and made it required reading for his colleagues in the late 1930s and early 1940s.[2]

In the same paper Turing reports another unexpected discovery: that of unsolvable problems. These are problems that are well defined with unique answers that can be shown to exist, but that we can also prove can never be computed by any Turing machine—that is to say, by *any* machine, a reversal of what had been a nineteenth-century dogma that problems that could be defined would ultimately be solved. Turing showed that there are as many unsolvable problems as solvable ones. Austrian American mathematician and philosopher Kurt Gödel reached a similar conclusion in his 1931 "incompleteness theorem." We are thus left with the perplexing situation of being able to define a problem, to prove that a unique answer exists, and yet know that the answer can never be found.

Turing had shown that at its essence, computation is based on a very simple mechanism. Because the Turing machine (and therefore any computer) is capable of basing its future course of action on results it has already computed, it is capable of making decisions and modeling arbitrarily complex hierarchies of information.

In 1939 Turing designed an electronic calculator called Bombe that helped decode messages that had been encrypted by the Nazi Enigma coding machine. By 1943, an engineering team influenced by Turing completed what is arguably the first computer, the Colossus, that enabled the Allies to continue decoding messages from more sophisticated versions of Enigma. The Bombe and Colossus were designed for a single task and could not be reprogrammed for a different one. But they performed this task brilliantly and are credited with having enabled the Allies to overcome the three-to-one advantage that the German Luftwaffe enjoyed over the British Royal Air Force and win the crucial Battle of Britain, as well as to continue anticipating Nazi tactics throughout the war.

It was on these foundations that John von Neumann created the architecture of the modern computer, which represents our third major idea.

Called the von Neumann machine, it has remained the core structure of essentially every computer for the past sixty-seven years, from the microcontroller in your washing machine to the largest supercomputers. In a paper dated June 30, 1945, and titled "First Draft of a Report on the EDVAC," von Neumann presented the ideas that have dominated computation ever since.[3] The von Neumann model includes a central processing unit, where arithmetical and logical operations are carried out; a memory unit, where the program and data are stored; mass storage; a program counter; and input/output channels. Although this paper was intended as an internal project document, it has become the bible for computer designers. You never know when a seemingly routine internal memo will end up revolutionizing the world.

The Turing machine was not designed to be practical. Turing's theorems were concerned not with the efficiency of solving problems but rather in examining the range of problems that could in theory be solved by computation. Von Neumann's goal, on the other hand, was to create a feasible concept of a computational machine. His model replaces Turing's one-bit computations with multiple-bit words (generally some multiple of eight bits). Turing's memory tape is sequential, so Turing machine programs spend an inordinate amount of time moving the tape back and forth to store and retrieve intermediate results. In contrast, von Neumann's memory is random access, so that any data item can be immediately retrieved.

One of von Neumann's key ideas is the stored program, which he had introduced a decade earlier: placing the program in the same type of random access memory as the data (and often in the same block of memory). This allows the computer to be reprogrammed for different tasks as well as for self-modifying code (if the program store is writable), which enables a powerful form of recursion. Up until that time, virtually all computers, including the Colossus, were built for a specific task. The stored program makes it possible for a computer to be truly universal, thereby fulfilling Turing's vision of the universality of computation.

Another key aspect of the von Neumann machine is that each instruction includes an operation code specifying the arithmetic or logical operation to be performed and the address of an operand from memory.

Von Neumann's concept of how a computer should be architected was introduced with his publication of the design of the EDVAC, a project he conducted with collaborators J. Presper Eckert and John Mauchly. The EDVAC itself did not actually run until 1951, by which time there were other stored-program computers, such as the Manchester Small-Scale Experimental Machine, ENIAC, EDSAC, and BINAC, all of which had been deeply influenced by von Neumann's paper and involved Eckert and Mauchly as designers. Von Neumann was a direct contributor to the design of a number of these machines, including a later version of ENIAC, which supported a stored program.

There were a few precursors to von Neumann's architecture, although with one surprising exception, none are true von Neumann machines. In 1944 Howard Aiken introduced the Mark I, which had an element of programmability but did not use a stored program. It read instructions from a punched paper tape and then executed each command immediately. It also lacked a conditional branch instruction.

In 1941 German scientist Konrad Zuse (1910–1995) created the Z-3 computer. It also read its program from a tape (in this case, coded on film) and also had no conditional branch instruction. Interestingly, Zuse had support from the German Aircraft Research Institute, which used the device to study wing flutter, but his proposal to the Nazi government for funding to replace his relays with vacuum tubes was turned down. The Nazis deemed computation as "not war important." That perspective goes a long way, in my view, toward explaining the outcome of the war.

There is actually one genuine forerunner to von Neumann's concept, and it comes from a full century earlier! English mathematician and inventor Charles Babbage's (1791–1871) Analytical Engine, which he first described in 1837, did incorporate von Neumann's ideas and featured a stored program via punched cards borrowed from the Jacquard loom.[4] Its random access memory included 1,000 words of 50 decimal digits each (the equivalent of about 21 kilobytes). Each instruction included an op code and an operand number, just like modern machine languages. It did include conditional branching and looping, so it was a true von Neumann machine. It was based entirely on mechanical gears and it appears that the Analytical Engine was beyond Babbage's design and organizational skills. He built parts of it but it never ran. It is unclear whether the twentieth-century pioneers of the computer, including von Neumann, were aware of Babbage's work.

Babbage's computer did result in the creation of the field of software programming. English writer Ada Byron (1815–1852), Countess of Lovelace and the only legitimate child of the poet Lord Byron, was the world's first computer programmer. She wrote programs for the Analytical Engine, which she needed to debug in her own mind (since the computer never worked), a practice well known to software engineers today as "table checking." She translated an article by the Italian mathematician Luigi Menabrea on the Analytical Engine and added extensive notes of her own, writing that "the Analytical Engine weaves algebraic patterns, just as the Jacquard loom weaves flowers and leaves." She went on to provide perhaps the first speculations on the feasibility of artificial intelligence, but concluded that the Analytical Engine has "no pretensions whatever to originate anything."

Babbage's conception is quite miraculous when you consider the era in which he lived and worked. However, by the mid-twentieth century, his ideas had been lost in the mists of time (although they were subsequently rediscovered). It was von Neumann who conceptualized and articulated the key principles of the computer as we know it today, and the world recognizes this by continuing to refer to the von Neumann machine as the principal model of computation. Keep in mind, though, that the von Neumann machine continually communicates data between its various units and within these units, so it could not be built without Shannon's theorems and the methods he devised for transmitting and storing reliable digital information.

That brings us to the fourth important idea, which is to go beyond Ada Byron's conclusion that a computer could not think creatively and find the key algorithms employed by the brain and then use these to turn a computer into a brain. Alan Turing introduced this goal in his 1950 paper "Computing Machinery and Intelligence," which includes his now-famous Turing test for ascertaining whether or not an AI has achieved a human level of intelligence.

In 1956 von Neumann began preparing a series of lectures intended for the prestigious Silliman lecture series at Yale University. Due to the ravages of cancer, he never delivered these talks nor did he complete the manuscript from which they were to be given. This unfinished document nonetheless remains a brilliant and prophetic foreshadowing of what I regard as humanity's most daunting and important project. It was published posthumously as *The Computer and the Brain* in 1958. It is fitting that the final work of one of the most brilliant mathematicians of the last century and one of the pioneers of the computer age was an examination of intelligence itself. This project was the earliest serious inquiry into the human brain from the perspective of a mathematician and computer scientist. Prior to von Neumann, the fields of computer science and neuroscience were two islands with no bridge between them.

Von Neumann starts his discussion by articulating the similarities and differences between the computer and the human brain. Given when he wrote this manuscript, it is remarkably accurate. He noted that the output of neurons was digital—an axon either fired or it didn't. This was far from obvious at the time, in that the output could have been an analog signal. The processing in the dendrites leading into a neuron and in the soma neuron cell body, however, was analog, and he described its calculations as a weighted sum of inputs with a threshold. This model of how neurons work led to the field of connectionism, which built systems based on this neuron model in both hardware and software. (As I described in the previous chapter, the first such connectionist system was created by Frank Rosenblatt as a software program on an IBM 704 computer at

Cornell in 1957, immediately after von Neumann's draft lectures became available.) We now have more sophisticated models of how neurons combine inputs, but the essential idea of analog processing of dendrite inputs using neurotransmitter concentrations has remained valid.

Von Neumann applied the concept of the universality of computation to conclude that even though the architecture and building blocks appear to be radically different between brain and computer, we can nonetheless conclude that a von Neumann machine can simulate the processing in a brain. The converse does not hold, however, because the brain is not a von Neumann machine and does not have a stored program as such (albeit we can simulate a very simple Turing machine in our heads). Its algorithm or methods are implicit in its structure. Von Neumann correctly concludes that neurons can learn patterns from their inputs, which we have now established are coded in part in dendrite strengths. What was not known in von Neumann's time is that learning also takes place through the creation and destruction of connections between neurons.

Von Neumann presciently notes that the speed of neural processing is extremely slow, on the order of a hundred calculations per second, but that the brain compensates for this through massive parallel processing— another unobvious and key insight. Von Neumann argued that each one of the brain's 10^{10} neurons (a tally that itself was reasonably accurate; estimates today are between 10^{10} and 10^{11}) was processing at the same time. In fact, each of the connections (with an average of about 10^3 to 10^4 connections per neuron) is computing simultaneously.

Von Neumann's estimates and his descriptions of neural processing are remarkable, given the primitive state of neuroscience at the time. One aspect of his work that I do disagree with, however, is his assessment of the brain's memory capacity. He assumes that the brain remembers every input for its entire life. Von Neumann assumes an average life span of 60 years, or about 2×10^9 seconds. With about 14 inputs to each neuron per second (which is actually low by at least three orders of magnitude) and with 10^{10} neurons, he arrives at an estimate of about 10^{20} bits for the brain's memory capacity. The reality, as I have noted earlier, is that we remember only a very small fraction of our thoughts and experiences, and even these memories are not stored as bit patterns at a low level (such as a video image), but rather as sequences of higher-level patterns.

As von Neumann describes each mechanism in the brain, he shows how a modern computer could accomplish the same thing, despite their apparent differences. The brain's analog mechanisms can be simulated through digital ones because digital computation can emulate analog values to any desired degree of precision (and the precision of analog information in the brain is quite low). The brain's massive parallelism can be simulated as well, given the significant speed advantage of computers in serial computation (an advantage that has vastly expanded over time). In addition, we can also use parallel processing in computers by using parallel von Neumann machines—which is exactly how supercomputers work today.

Von Neumann concludes that the brain's methods cannot involve lengthy sequential algorithms, when one considers how quickly humans are able to make decisions combined with the very slow computational speed of neurons. When a third baseman fields a ball and decides to throw to first rather than to second base, he makes this decision in a fraction of a second, which is only enough time for each neuron to go through a handful of cycles. Von Neumann concludes correctly that the brain's remarkable powers come from all its 100 billion neurons being able to process information simultaneously. As I have noted, the visual cortex makes sophisticated visual judgments in only three or four neural cycles.

There is considerable plasticity in the brain, which enables us to learn. But there is far greater plasticity in a computer, which can completely restructure its methods by changing its software. Thus, in that respect, a computer will be able to emulate the brain, but the converse is not the case.

When von Neumann compared the capacity of the brain's massively parallel organization to the (few) computers of his time, it was clear that the brain had far greater memory and speed. By now the first supercomputer to achieve specifications matching some of the more conservative estimates of the speed required to functionally simulate the human brain (about 10^{16} operations per second) has been built.[5] (I estimate that this level of computation will cost $1,000 by the early 2020s.) With regard to memory we are even closer. Even though it was remarkably early in the history of the computer when his manuscript was written, von Neumann nonetheless had confidence that both the hardware and software of human intelligence would ultimately fall into place, which was his motivation for having prepared these lectures.

Von Neumann was deeply aware of the increasing pace of progress and its profound implications for humanity's future. A year after his death in 1957, fellow mathematician Stan Ulam quoted him as having said in the early 1950s that "the ever accelerating progress of technology and changes in the mode of human life give the appearance of approaching some essential singularity in the history of the race beyond which human affairs, as we know them, could not continue." This is the first known use of the word "singularity" in the context of human technological history.

Von Neumann's fundamental insight was that there is an essential equivalence between a computer and the brain. Note that the emotional intelligence of a biological human is part of its intelligence. If von Neumann's insight is correct, and if one accepts my own leap of faith that a nonbiological entity that convincingly re-creates the intelligence (emotional and otherwise) of a biological human is conscious (see the next chapter), then one would have to conclude that there is an essential equivalence between a computer—*with the right software*—and a (conscious) mind. So is von Neumann correct?

Most computers today are entirely digital, whereas the human brain combines digital and analog methods. But analog methods are easily and routinely re-created by digital ones to any desired level of accuracy. American computer scientist Carver Mead (born in 1934) has shown that we can directly emulate the brain's analog methods in silicon, which he has demonstrated with what he calls "neuromorphic" chips.[6] Mead has demonstrated how this approach can be thousands of times more efficient than digitally emulating analog methods. As we codify the massively repeated neocortical algorithm, it will make sense to use Mead's approach. The IBM Cognitive Computing Group, led by Dharmendra Modha, has introduced chips that emulate neurons and their connections, including the ability to form new connections.[7] Called "SyNAPSE," one of the chips provides a direct simulation of 256 neurons with about a quarter million synaptic connections. The goal of the project is to create a simulated neocortex with 10 billion neurons and 100 trillion connections—close to a human brain—that uses only one kilowatt of power.

As von Neumann described over a half century ago, the brain is extremely slow but massively parallel. Today's digital circuits are at least 10 million times faster than the brain's electrochemical switches. Conversely, all 300 million of the brain's neocortical pattern recognizers process simultaneously, and all quadrillion of its interneuronal connections are potentially computing at the same time. The key issue for providing the requisite hardware to successfully model a human brain, though, is the overall memory and computational throughput required. We do not need to directly copy the brain's architecture, which would be a very inefficient and inflexible approach.

Let's estimate what those hardware requirements are. Many projects have attempted to emulate the type of hierarchical learning and pattern recognition that takes place in the neocortical hierarchy, including my own work with hierarchical hidden Markov models. A conservative estimate from my own experience is that emulating one cycle in a single pattern recognizer in the biological brain's neocortex would require about 3,000 calculations. Most simulations run at a fraction of this estimate. With the brain running at about 10^2 (100) cycles per second, that comes to 3×10^5 (300,000) calculations per second per pattern recognizer. Using my estimate of 3×10^8 (300 million) pattern recognizers, we get about 10^{14} (100 trillion) calculations per second, a figure that is consistent with my estimate in *The Singularity Is Near*. In that book I projected that to functionally simulate the brain would require between 10^{14} and 10^{16} calculations per second (cps) and used 10^{16} cps to be conservative. AI expert Hans Moravec's estimate, based on extrapolating the computational requirement of the early (initial) visual processing across the entire brain, is 10^{14} cps, which matches my own assessment here.

Routine desktop machines can reach 10^{10} cps, although this level of performance can be significantly amplified by using cloud resources. The fastest supercomputer, Japan's K Computer, has already reached 10^{16} cps.[8] Given that the algorithm of the neocortex is massively repeated, the approach of using neuromorphic chips such as the IBM SyNAPSE chips mentioned above is also promising.

In terms of memory requirement, we need about 30 bits (about four bytes) for one connection to address one of 300 million other pattern recognizers. If we estimate an average of eight inputs to each pattern recognizer, that comes to 32 bytes per recognizer. If we add a one-byte weight for each input, that brings us to 40 bytes. Add another 32 bytes for downward connections, and we are at 72 bytes. Note that the branching-up-and-down figure will often be much higher than eight, though these very large branching trees are shared by many recognizers. For example, there may be hundreds of recognizers involved in recognizing the letter "p." These will feed up into thousands of such recognizers at this next higher level that deal with words and phrases that include "p." However, each "p" recognizer does not repeat the tree of connections that feeds up to all of the words and phrases that include "p"—they all share one such tree of connections. The same is true of downward connections: A recognizer that is responsible for the word "APPLE" will tell all of the thousands of "E" recognizers at a level below it that an "E" is expected if it has already seen "A," "P," "P," and "L." That tree of connections is not repeated for each word or phrase recognizer that wants to inform the next lower level that an "E" is expected. Again, they are shared. For this reason, an overall estimate of eight up and eight down on average per pattern recognizer is reasonable. Even if we increase this particular estimate, it does not significantly change the order of magnitude of the resulting estimate.

With 3×10^8 (300 million) pattern recognizers at 72 bytes each, we get an overall memory requirement of about 2×10^{10} (20 billion) bytes. That is actually a quite modest number that routine computers today can exceed.

These estimates are intended only to provide rough estimates of the order of magnitude required. Given that digital circuits are inherently about 10 million times faster than the biological neocortical circuits, we do not need to match the human brain for parallelism—modest parallel processing (compared with the trillions-fold parallelism of the human brain) will be sufficient. We can see that the necessary computational requirements are coming within reach. The brain's rewiring of itself—dendrites are continually creating new synapses—can also be emulated in software using links, a far more flexible system than the brain's method of plasticity, which as we have seen is impressive but limited.

The redundancy used by the brain to achieve robust invariant results can certainly be replicated in software emulations. The mathematics of optimizing these types of self-organizing hierarchical learning systems is well understood. The organization of the brain is far from optimal. Of course it didn't need to be—it only needed to be good enough to achieve the threshold of being able to create tools that would compensate for its own limitations.

Another restriction of the human neocortex is that there is no process that eliminates or even reviews contradictory ideas, which accounts for why human thinking is often massively inconsistent. We have a weak mechanism to address this called critical thinking, but this skill is not practiced nearly as often as it should be. In a software-based neocortex, we can build in a process that reveals inconsistencies for further review.

It is important to note that the design of an entire brain region is simpler than the design of a single neuron. As discussed earlier, models often get simpler at a higher level—consider an analogy with a computer. We do need to understand the detailed physics of semiconductors to model a transistor, and the equations underlying a single real transistor are complex. A digital circuit that multiples two numbers requires hundreds of them. Yet we can model this multiplication circuit very simply with one or two formulas. An entire computer with billions of transistors can be modeled through its instruction set and register description, which can be described on a handful of written pages of text and formulas. The software programs for an operating system, language compilers, and assemblers are reasonably complex, but modeling a particular program—for example, a speech recognition program based on hierarchical hidden Markov modeling—may likewise be described in only a few pages of equations. Nowhere in such a description would be found the details of semiconductor physics or even of computer architecture.

A similar observation holds true for the brain. A particular neocortical pattern recognizer that detects a particular invariant visual feature (such as a face) or that performs a bandpass filtering (restricting input to a specific frequency range) on sound or that evaluates the temporal proximity of two events can be described with far fewer specific details than the actual physics and chemical relations controlling the neurotransmitters, ion channels, and other synaptic and dendritic variables involved in the neural processes. Although all of this complexity needs to be carefully considered before advancing to the next higher conceptual level, much of it can be simplified as the operating principles of the brain are revealed.

CHAPTER 9

THOUGHT EXPERIMENTS ON THE MIND

Minds are simply what brains do.

—Marvin Minsky, *The Society of Mind*

When intelligent machines are constructed, we should not be surprised to find them as confused and as stubborn as men in their convictions about mind-matter, consciousness, free will, and the like.

—Marvin Minsky, *The Society of Mind*

Who Is Conscious?

The real history of consciousness starts with one's first lie.

—Joseph Brodsky

Suffering is the sole origin of consciousness.

—Fyodor Dostoevsky, *Notes from Underground*

There is a kind of plant that eats organic food with its flowers: when a fly settles upon the blossom, the petals close upon it and hold it fast till the plant has absorbed the insect into its system; but they will close on nothing but what is good to eat; of a drop of rain or a piece of stick they will take no notice. Curious! that so unconscious a thing should have such a keen eye to its own interest. If this is unconsciousness, where is the use of consciousness?

—Samuel Butler, 1871

We have been examining the brain as an entity that is capable of certain levels of accomplishment. But that perspective essentially leaves our *selves* out of the picture. We appear to live in our brains. We have subjective lives. How does the objective view of the brain that we have discussed up until now relate to our own feelings, to our sense of being the person having the experiences?

British philosopher Colin McGinn (born in 1950) writes that discussing "consciousness can reduce even the most fastidious thinker to blabbering incoherence." The reason for this is that people often have unexamined and inconsistent views on exactly what the term means.

Many observers consider consciousness to be a form of performance— for example, the capacity for self-reflection, that is, the ability to understand one's own thoughts and to explain them. I would describe that as the ability to think about one's own thinking. Presumably, we could come up with a way of evaluating this ability and then use this test to separate conscious things from unconscious things.

However, we quickly get into trouble in trying to implement this approach. Is a baby conscious? A dog? They're not very good at describing their own thinking process. There are people who believe that babies and dogs are not conscious beings precisely because they cannot explain themselves. How about the computer known as Watson? It can be put into a mode where it actually does explain how it came up with a given answer. Because it contains a model of its own thinking, is Watson therefore conscious whereas the baby and the dog are not?

Before we proceed to parse this question further, it is important to reflect on the most significant distinction relating to it: What is it that we can ascertain from science, versus what remains truly a matter of philosophy? One view is that philosophy is a kind of halfway house for questions that have not yet yielded to the scientific method. According to this perspective, once science advances sufficiently to resolve a particular set of questions, philosophers can then move on to other concerns, until such time that science resolves them also. This view is endemic where the issue of consciousness is concerned, and specifically the question "What and who is conscious?"

Consider these statements by philosopher John Searle: "We know that brains cause consciousness with specific biological mechanisms.... The essential thing is to recognize that consciousness is a biological process like digestion, lactation, photosynthesis, or mitosis.... The brain is a machine, a biological machine to be sure, but a machine all the same. So the first step is to figure out how the brain does it and then build an artificial machine that has an equally effective mechanism for causing consciousness."[1] People are often surprised to see these quotations because they assume that Searle is devoted to protecting the mystery of consciousness against reductionists like Ray Kurzweil.

The Australian philosopher David Chalmers (born in 1966) has coined the term "the hard problem of consciousness" to describe the difficulty of pinning down this essentially indescribable concept. Sometimes a brief phrase encapsulates an entire school of thought so well that it becomes emblematic (for example, Hannah Arendt's "the banality of evil"). Chalmers's famous formulation accomplishes this very well.

When discussing consciousness, it becomes very easy to slip into considering the observable and measurable attributes that we associate with being conscious, but this approach misses the very essence of the idea. I just mentioned the concept of metacognition—the idea of thinking about one's own thinking—as one such correlate of consciousness. Other observers conflate emotional intelligence or moral intelligence with consciousness. But, again, our ability to express a loving sentiment, to get the joke, or to be sexy are simply types of performances—impressive and intelligent perhaps, but skills that can nonetheless be observed and measured (even if we argue about how to assess them). Figuring out how the brain accomplishes these sorts of tasks and what is going on in the brain when we do them constitutes Chalmers's "easy" question of consciousness. Of course, the "easy" problem is anything but and represents perhaps the most difficult and important scientific quest of our era. Chalmers's "hard" question, meanwhile, is so hard that it is essentially ineffable.

In support of this distinction, Chalmers introduces a thought experiment involving what he calls zombies. A zombie is an entity that acts just like a person but simply does not have subjective experience—that is, a zombie is not conscious. Chalmers argues that since we can conceive of zombies, they are at least logically possible. If you were at a cocktail party and there were both "normal" humans and zombies, how would you tell the difference? Perhaps this sounds like a cocktail party you have attended.

Many people answer this question by saying they would interrogate individuals they wished to assess about their emotional reactions to events and ideas. A zombie, they believe, would betray its lack of subjective experience through a deficiency in certain types of emotional responses. But an answer along these lines simply fails to appreciate the assumptions of the thought experiment. If we encountered an unemotional person (such as an individual with certain emotional deficits, as is common in certain types of autism) or an avatar or a robot that was not convincing as an emotional human being, then that entity is not a zombie. Remember: According to Chalmers's assumption, a zombie *is* completely normal in his ability to respond, including the ability to react emotionally; he is just lacking subjective experience. The bottom line is that there is no way to identify a zombie, because by definition there is no apparent indication of his zombie nature in his behavior. So is this a distinction without a difference?

Chalmers does not attempt to answer the hard question but does provide some possibilities. One is a form of dualism in which consciousness per se does not exist in the physical world but rather as a separate ontological reality. According to this formulation, what a person does is based on the processes in her brain. Because the brain is causally closed, we can fully explain a person's actions, including her thoughts, through its processes. Consciousness then exists essentially in another realm, or at least is a property separate from the physical world. This explanation does not permit the mind (that is to say, the conscious property associated with the brain) to causally affect the brain.

Another possibility that Chalmers entertains, which is not logically distinct from his notion of dualism, and is often called panprotopsychism, holds that all physical systems are conscious, albeit a human is more conscious than, say, a light switch. I would certainly agree that a human brain has more to be conscious about than a light switch.

My own view, which is perhaps a subschool of panprotopsychism, is that consciousness is an emergent property of a complex physical system. In this view a dog is also conscious but somewhat less than a human. An ant has some level of consciousness, too, but much less that of a dog. The ant colony, on the other hand, could be considered to have a higher level of consciousness than the individual ant; it is certainly more intelligent than a lone ant. By this reckoning, a computer that is successfully emulating the complexity of a human brain would also have the same emergent consciousness as a human.

Another way to conceptualize the concept of consciousness is as a system that has "qualia." So what are qualia? One definition of the term is "conscious experiences." That, however, does not take us very far. Consider this thought experiment: A neuroscientist is completely color-blind—not the sort of color-blind in which one mixes up certain shades of, say, green and red (as I do), but rather a condition in which the afflicted individual lives entirely in a black-and-white world. (In a more extreme version of this scenario, she has grown up in a black-and-white world and has never seen any colors. Bottom line, there is no color in her world.) However, she has extensively studied the physics of color—she is aware that the wavelength of red light is 700 nanometers—as well as the neurological processes of a person who can experience colors normally, and thus knows a great deal about how the brain processes color. She knows more about color than most people. If you wanted to help her out and explain what this actual experience of "red" is like, how would you do it?

Perhaps you would read her a section from the poem "Red" by the Nigerian poet Oluseyi Oluseun:

Red the colour of blood
the symbol of life
Red the colour of danger
the symbol of death

Red the colour of roses
the symbol of beauty
Red the colour of lovers
the symbol of unity

Red the colour of tomato
the symbol of good health
Red the colour of hot fire
the symbol of burning desire

That actually would give her a pretty good idea of some of the associations people have made with red, and may even enable her to hold her own in a conversation about the color. ("Yes, I love the color red, it's so hot and fiery, so dangerously beautiful . . .") If she wanted to, she could probably convince people that she had experienced red, but all the poetry in the world would not actually enable her to have that experience.

Similarly, how would you explain what it feels like to dive into water to someone who has never touched water? We would again be forced to resort to poetry, but there is really no way to impart the experience itself. These experiences are what we refer to as qualia.

Many of the readers of this book have experienced the color red. But how do I know whether your experience of red is not the same experience that I have when I look at blue? We both look at a red object and state assuredly that it is red, but that does not answer the question. I may be experiencing what you experience when you look at blue, but we have both learned to call red things red. We could start swapping poems again, but they would simply reflect the associations that people have made with colors; they do not speak to the actual nature of the qualia. Indeed, congenitally blind people have read a great deal about colors, as such references are replete in literature, and thus they do have some version of an experience of color. How does their experience of red compare with the experience of sighted people? This is really the same question as the one concerning the woman in the black-and-white world. It is remarkable that such common phenomena in our lives are so completely ineffable as to make a simple confirmation, like one that we are experiencing the same qualia, impossible.

Another definition of qualia is the feeling of an experience. However, this definition is no less circular than our attempts at defining consciousness above, as the phrases "feeling," "having an experience," and "consciousness" are all synonyms. Consciousness and the closely related question of qualia are a fundamental, perhaps the ultimate, philosophical question (although the issue of identity may be even more important, as I will discuss in the closing section of this chapter).

So with regard to consciousness, what exactly *is* the question again? It is this: Who or what is conscious? I refer to "mind" in the title of this book rather than "brain" because a mind is a brain that is conscious. We could also say that a mind has free will and identity. The assertion that these issues are philosophical is itself not self-evident. I maintain that these questions can never be fully resolved through science. In other words, there are no falsifiable experiments that we can contemplate that would resolve them, not without making philosophical assumptions. If we were building a consciousness detector, Searle would want it to ascertain that it was squirting biological neurotransmitters. American philosopher Daniel Dennett (born in 1942) would be more flexible on substrate, but might want to determine whether or not the system contained a model of itself and of its own performance. That view comes closer to my own, but at its core is still a philosophical assumption.

Proposals have been regularly presented that purport to be scientific theories linking consciousness to some measurable physical attribute— what Searle refers to as the "mechanism for causing consciousness." American scientist, philosopher, and anesthesiologist Stuart Hameroff (born in 1947) has written that "cytoskeletal filaments are the roots of consciousness."[2] He is referring to thin threads in every cell (including neurons but not limited to them) called microtubules, which give each cell structural integrity and play a role in cell division. His books and papers on this issue contain detailed descriptions and equations that explain the plausibility that the microtubules play a role in information processing within the cell. But the connection of microtubules to consciousness requires a leap of faith not fundamentally different from the leap of faith implicit in a religious doctrine that describes a supreme being bestowing consciousness (sometimes referred to as a "soul") to certain (usually human) entities. Some weak evidence is proffered for Hameroff's view, specifically the observation that the neurological processes that could support this purported cellular computing are stopped during anesthesia. But this is far from compelling substantiation, given that lots of processes are halted during anesthesia. We cannot even say for certain that subjects are not conscious when anesthetized. All we do know is that people do not remember their experiences afterward. Even that is not universal, as some people do remember—accurately—their experience while under anesthesia, including, for example, conversations by their surgeons. Called anesthesia awareness, this phenomenon is estimated to occur about 40,000 times a year in the United States.[3] But even setting that aside, consciousness and memory are completely different concepts. As I have discussed extensively, if I think back on my moment-to-moment experiences over the past day, I have had a vast number of sensory impressions yet I remember very few of them.

Was I therefore not conscious of what I was seeing and hearing all day? It is actually a good question, and the answer is not so clear.

English physicist and mathematician Roger Penrose (born in 1931) took a different leap of faith in proposing the source of consciousness, though his also concerned the microtubules—specifically, their purported quantum computing abilities. His reasoning, although not explicitly stated, seemed to be that consciousness is mysterious, and a quantum event is also mysterious, so they must be linked in some way.

Penrose started his analysis with Turing's theorems on unsolvable problems and Gödel's related incompleteness theorem. Turing's premise (which was discussed in greater detail in chapter 8) is that there are algorithmic problems that can be stated but that cannot be solved by a Turing machine. Given the computational universality of the Turing machine, we can conclude that these "unsolvable problems" cannot be solved by any machine. Gödel's incompleteness theorem has a similar result with regard to the ability to prove conjectures involving numbers. Penrose's argument is that the human brain is able to solve these unsolvable problems, so is therefore capable of doing things that a deterministic machine such as a computer is unable to do. His motivation, at least in part, is to elevate human beings above machines. But his central premise—that humans can solve Turing's and Gödel's insoluble problems—is unfortunately simply not true.

A famous unsolvable problem called the busy beaver problem is stated as follows: Find the maximum number of 1s that a Turing machine with a certain number of states can write on its tape. So to determine the busy beaver of the number n, we build all of the Turing machines that have n states (which will be a finite number if n is finite) and then determine the largest number of 1s that these machines write on their tapes, excluding those Turing machines that get into an infinite loop. This is unsolvable because as we seek to simulate all of these n-state Turing machines, our simulator will get into an infinite loop when it attempts to simulate one of the Turing machines that does get into an infinite loop. However, it turns out that computers have nonetheless been able to determine the busy beaver function for certain ns. So have humans, but computers have solved the problem for far more ns than unassisted humans. Computers are generally better than humans at solving Turing's and Gödel's unsolvable problems.

Penrose linked these claimed transcendent capabilities of the human brain to the quantum computing that he hypothesized took place in it. According to Penrose, these neural quantum effects were somehow inherently not achievable by computers, so therefore human thinking has an inherent edge. In fact, common electronics uses quantum effects (transistors rely on quantum tunneling of electrons across barriers); quantum computing in the brain has never been demonstrated; human mental performance can be satisfactorily explained by classical computing methods; and in any event nothing bars us from applying quantum computing in computers. None of these objections has ever been addressed by Penrose. It was when critics pointed out that the brain is a warm and messy place for quantum computing that Hameroff and Penrose joined forces. Penrose found a perfect vehicle within neurons that could conceivably support quantum computing—namely, the microtubules that Hameroff had speculated were part of the information processing within a neuron. So the Hameroff-Penrose thesis is that the microtubules in the neurons are doing quantum computing and that this is responsible for consciousness.

This thesis has also been criticized, for example, by Swedish American physicist and cosmologist Max Tegmark (born in 1967), who determined that quantum events in microtubules could survive for only 10^{-13} seconds, which is much too brief a period of time either to compute results of any significance or to affect neural processes. There are certain types of problems for which quantum computing would show superior capabilities to classical computing—for example, the cracking of encryption codes through the factoring of large numbers. However, unassisted human thinking has proven to be terrible at solving them, and cannot match even classical computers in this area, which suggests that the brain is not demonstrating any quantum computing capabilities. Moreover, even if such a phenomenon as quantum computing in the brain did exist, it would not necessarily be linked to consciousness.

You Gotta Have Faith

> What a piece of work is a man! How noble in reason! How infinite in faculties! In form and moving, how express and admirable! In action how like an angel! In apprehension, how like a god! The beauty of the world! The paragon of animals! And yet, to me, what is this quintessence of dust?
>
> —Hamlet, in Shakespeare's *Hamlet*

The reality is that these theories are all leaps of faith, and I would add that where consciousness is concerned, the guiding principle is "you gotta have faith"—that is, we each need a leap of faith as to what and who is conscious, and who and what we are as conscious beings. Otherwise we could not get up in the morning. But we should be honest about the fundamental need for a leap of faith in this matter and self-reflective as to what our own particular leap involves.

People have very different leaps, despite impressions to the contrary. Individual philosophical assumptions about the nature and source of consciousness underlie disagreements on issues ranging from animal rights to abortion, and will result in even more contentious future conflicts over machine rights. My objective prediction is that machines in the future will appear to be conscious and that they will be convincing to biological people when they speak of their qualia. They will exhibit the full range of subtle, familiar emotional cues; they will make us laugh and cry; and they will get mad at us if we say that we don't believe that they are conscious. (They will be very smart, so we won't want that to happen.) We will come to accept that they are conscious persons. My own leap of faith is this: Once machines do succeed in being convincing when they speak of their qualia and conscious experiences, they will indeed constitute conscious persons. I have come to my position via this thought experiment: Imagine that you meet an entity in the future (a robot or an avatar) that is completely convincing in her emotional reactions. She laughs convincingly at your jokes, and in turn makes you laugh and cry (but not just by pinching you). She convinces you of her sincerity when she speaks of her fears and longings. In every way, she seems conscious. She seems, in fact, like a person. Would you accept her as a conscious person?

If your initial reaction is that you would likely detect some way in which she betrays her nonbiological nature, then you are not keeping to the assumptions in this hypothetical situation, which established that she *is* fully convincing. Given that assumption, if she were threatened with destruction and responded, as a human would, with terror, would you react in the same empathetic way that you would if you witnessed such a scene involving a human? For myself, the answer is yes, and I believe the answer would be the same for most if not virtually all other people regardless of what they might assert now in a philosophical debate. Again, the emphasis here is on the word "convincing."

There is certainly disagreement on when or even whether we will encounter such a nonbiological entity. My own consistent prediction is that this will first take place in 2029 and become routine in the 2030s. But putting the time frame aside, I believe that we will eventually come to regard such entities as conscious. Consider how we already treat them when we are exposed to them as characters in stories and movies: R2D2 from the *Star Wars* movies, David and Teddy from the movie *A.I.,* Data from the TV series *Star Trek: The Next Generation*, Johnny 5 from the movie *Short Circuit,* WALL-E from Disney's movie *Wall-E,* T-800—the (good) Terminator—in the second and later *Terminator* movies, Rachael the Replicant from the movie *Blade Runner* (who, by the way, is not aware that she is not human), Bumblebee from the movie, TV, and comic series *Transformers,* and Sonny from the movie *I, Robot.* We do empathize with these characters even though we know that they are nonbiological. We regard them as conscious persons, just as we do biological human characters. We share their feelings and fear for them when they get into trouble. If that is how we treat fictional nonbiological characters today, then that is how we will treat real-life intelligences in the future that don't happen to have a biological substrate.

If you do accept the leap of faith that a nonbiological entity that is convincing in its reactions to qualia is actually conscious, then consider what that implies: namely that consciousness is an emergent property of the overall pattern of an entity, not the substrate it runs on.

There is a conceptual gap between science, which stands for *objective* measurement and the conclusions we can draw thereby, and consciousness, which is a synonym for *subjective* experience. We obviously cannot simply ask an entity in question, "Are you conscious?" If we look inside its "head," biological or otherwise, to ascertain that, then we would have to make philosophical assumptions in determining what it is that we are looking for. The question as to whether or not an entity is conscious is therefore not a scientific one. Based on this, some observers go on to question whether consciousness itself has any basis in reality. English writer and philosopher Susan Blackmore (born in 1951) speaks of the "grand illusion of consciousness." She acknowledges the reality of the meme (idea) of consciousness—in other words, consciousness certainly exists as an idea, and there are a great many neocortical structures that deal with the idea, not to mention words that have been spoken and written about it. But it is not clear that it refers to something real. Blackburn goes on to explain that she is not necessarily denying the reality of consciousness, but rather attempting to articulate the sorts of dilemmas we encounter when we try to pin down the concept. As British psychologist and writer Stuart Sutherland (1927–1998) wrote in the *International Dictionary of Psychology,* "Consciousness is a fascinating but elusive phenomenon; it is impossible to specify what it is, what it does, or why it evolved."[4]

However, we would be well advised not to dismiss the concept too easily as just a polite debate between philosophers—which, incidentally, dates back two thousand years to the Platonic dialogues. The idea of consciousness underlies our moral system, and our legal system in turn is loosely built on those moral beliefs. If a person extinguishes someone's consciousness, as in the act of murder, we consider that to be immoral, and with some exceptions, a high crime. Those exceptions are also relevant to consciousness, in that we might authorize police or military forces to kill certain conscious people to protect a greater number of other conscious people. We can debate the merits of particular exceptions, but the underlying principle holds true.

Assaulting someone and causing her to experience suffering is also generally considered immoral and illegal. If I destroy my property, it is probably acceptable. If I destroy your property without your permission, it is probably not acceptable, but not because I am causing suffering to your property, but rather to you as the owner of the property. On the other hand, if my property includes a conscious being such as an animal, then I as the owner of that animal do not necessarily have free moral or legal rein to do with it as I wish—there are, for example, laws against animal cruelty.

Because a great deal of our moral and legal system is based on protecting the existence of and preventing the unnecessary suffering of conscious entities, in order to make responsible judgments we need to answer the question as to who is conscious. That question is therefore not simply a matter for intellectual debate, as is evident in the controversy surrounding an issue like abortion. I should point out that the abortion issue can go somewhat beyond the issue of consciousness, as pro-life proponents argue that the potential for an embryo to ultimately become a conscious person is sufficient reason for it to be awarded protection, just as someone in a coma deserves that right. But fundamentally the issue is a debate about when a fetus becomes conscious.

Perceptions of consciousness also often affect our judgments in controversial areas. Looking at the abortion issue again, many people make a distinction between a measure like the morning-after pill, which prevents the implantation of an embryo in the uterus in the first days of pregnancy, and a late-stage abortion. The difference has to do with the likelihood that the late-stage fetus is conscious. It is difficult to maintain that a few-days-old embryo is conscious unless one takes a panprotopsychist position, but even in these terms it would rank below the simplest animal in terms of consciousness. Similarly, we have very different reactions to the maltreatment of great apes versus, say, insects. No one worries too much today about causing pain and suffering to our computer software (although we do comment extensively on the ability of software to cause us suffering), but when future software has the intellectual, emotional, and moral intelligence of biological humans, this will become a genuine concern.

Thus my position is that I will accept nonbiological entities that are fully convincing in their emotional reactions to be conscious persons, and my prediction is that the consensus in society will accept them as well. Note that this definition extends beyond entities that can pass the Turing test, which requires mastery of human language. The latter are sufficiently humanlike that I would include them, and I believe that most of society will as well, but I also include entities that evidence humanlike emotional reactions but may not be able to pass the Turing test—for example, young children.

Does this resolve the philosophical question of who is conscious, at least for myself and others who accept this particular leap of faith? The answer is: *not quite*. We've only covered one case, which is that of entities that act in a humanlike way. Even though we are discussing future entities that are not biological, we are talking about entities that demonstrate convincing humanlike reactions, so this position is still human-centric. But what about more alien forms of intelligence that are not humanlike? We can imagine intelligences that are as complex as or perhaps vastly more complex and intricate than human brains, but that have completely different emotions and motivations. How do we decide whether or not they are conscious?

We can start by considering creatures in the biological world that have brains comparable to those of humans yet evince very different sorts of behaviors. British philosopher David Cockburn (born in 1949) writes about viewing a video of a giant squid that was under attack (or at least it thought it was—Cockburn hypothesized that it might have been afraid of the human with the video camera). The squid shuddered and cowered, and Cockburn writes, "It responded in a way which struck me immediately and powerfully as one of fear. Part of what was striking in this sequence was the way in which it was possible to see in the behavior of a creature physically so very different from human beings an emotion which was so unambiguously and specifically one of fear."[5] He concludes that the animal was feeling that emotion and he articulates the belief that most other people viewing that film would come to the same conclusion. If we accept Cockburn's description and conclusion, then we would have to add giant squids to our list of conscious entities. However, this has not gotten us very far either, because it is still based on our empathetic reaction to an emotion that we recognize in ourselves. It is still a self-centric or human-centric perspective.

If we step outside biology, nonbiological intelligence will be even more varied than intelligence in the biological world. For example, some entities may not have a fear of their own destruction, and may not have a need for the emotions we see in humans or in any biological creature. Perhaps they could still pass the Turing test, or perhaps they wouldn't even be willing to try.

We do in fact build robots today that do not have a sense of self-preservation to carry out missions in dangerous environments. They're not sufficiently intelligent or complex yet for us to seriously consider their sentience, but we can imagine future robots of this sort that are as complex as humans. What about them?

Personally I would say that if I saw in such a device's behavior a commitment to a complex and worthy goal and the ability to execute notable decisions and actions to carry out its mission, I would be impressed and probably become upset if it got destroyed. This is now perhaps stretching the concept a bit, in that I am responding to behavior that does not include many emotions we consider universal in people and even in biological creatures of all kinds. But again, I am seeking to connect with attributes that I can relate to in myself and other people. The idea of an entity totally dedicated to a noble goal and carrying it out or at least attempting to do so without regard for its own well-being is, after all, not completely foreign to human experience. In this instance we are also considering an entity that is seeking to protect biological humans or in some way advance our agenda.

What if this entity has its own goals distinct from a human one and is not carrying out a mission we would recognize as noble in our own terms? I might then attempt to see if I could connect or appreciate some of its abilities in some other way. If it is indeed very intelligent, it is likely to be good at math, so perhaps I could have a conversation with it on that topic. Maybe it would appreciate math jokes.

But if the entity has no interest in communicating with me, and I don't have sufficient access to its actions and decision making to be moved by the beauty of its internal processes, does that mean that it is not conscious? I need to conclude that entities that do not succeed in convincing me of their emotional reactions, or that don't care to try, are not necessarily not conscious. It would be difficult to recognize another conscious entity without establishing some level of empathetic communication, but that judgment reflects my own limitations more than it does the entity under consideration. We thus need to proceed with humility. It is challenging enough to put ourselves in the subjective shoes of another human, so the task will be that much harder with intelligences that are extremely different from our own.

What Are We Conscious Of?

If we could look through the skull into the brain of a consciously thinking person, and if the place of optimal excitability were luminous, then we should see playing over the cerebral surface, a bright spot with fantastic, waving borders constantly fluctuating in size and form, surrounded by a darkness more or less deep, covering the rest of the hemisphere.

—Ivan Petrovich Pavlov, 1913[6]

Returning to the giant squid, we can recognize some of its apparent emotions, but much of its behavior is a mystery. What is it like being a giant squid? How does it feel as it squeezes its spineless body through a tiny opening? We don't even have the vocabulary to answer this question, given that we cannot even describe experiences that we do share with other people, such as seeing the color red or feeling water splash on our bodies.

But we don't have to go as far as the bottom of the ocean to find mysteries in the nature of conscious experiences—we need only consider our own. I know, for example, that I am conscious. I assume that you, the reader, are conscious also. (As for people who have not bought my book, I am not so sure.) But what am I conscious *of*? You might ask yourself the same question.

Try this thought experiment (which will work for those of you who drive a car): Imagine that you are driving in the left lane of a highway. Now close your eyes, grab an imagined steering wheel, and make the movements to change lanes to the lane to your right.

Okay, before continuing to read, try it.

Here is what you probably did: You held the steering wheel. You checked that the right lane is clear. Assuming the lane was clear, you turned the steering wheel to the right for a brief period. Then you straightened it out again. Job done.

It's a good thing you weren't in a real car, because you just zoomed across all the lanes of the highway and crashed into a tree. While I probably should have mentioned that you shouldn't try this in a real moving car (but then I assume you have already mastered the rule that you shouldn't drive with your eyes closed), that's not really the key problem here. If you used the procedure I just described—and almost everyone does when doing this thought experiment—you got it wrong. Turning the wheel to the right and then straightening it out causes the car to head in a direction that is diagonal to its original direction. It will cross the lane to the right, as you intended, but it will keep going to the right indefinitely until it zooms off the road. What you needed to do as your car crossed the lane to the right was to then turn the wheel to the left, just as far as you had turned it to the right, and *then* straighten it out again. This will cause the car to again head straight in the new lane.

Consider the fact that if you're a regular driver, you've done this maneuver thousands of times. Are you not conscious when you do this? Have you never paid attention to what you are actually doing when you change lanes? Assuming that you are not reading this book in a hospital while recovering from a lane-changing accident, you have clearly mastered this skill. Yet you are not conscious of what you did, however many times you've accomplished this task.

When people tell stories of their experiences, they describe them as sequences of situations and decisions. But this is not how we experience a story in the first place. Our original experience is as a sequence of high-level patterns, some of which may have triggered feelings. We remember only a small subset of those patterns, if that. Even if we are reasonably accurate in our recounting of a story, we use our powers of confabulation to fill in missing details and convert the sequence into a coherent tale. We cannot be certain what our original conscious experience was from our recollection of it, yet memory is the only access we have to that experience. The present moment is, well, fleeting, and is quickly turned into a memory, or, more often, not. Even if an experience is turned into a memory, it is stored, as the PRTM indicates, as a high-level pattern composed of other patterns in a huge hierarchy. As I have pointed out several times, almost all of the experiences we have (like any of the times we changed lanes) are immediately forgotten. So ascertaining what constitutes our own conscious experience is actually not attainable.

East Is East and West Is West

> Before brains there was no color or sound in the universe, nor was there any flavor or aroma and probably little sense and no feeling or emotion.
>
> —Roger W. Sperry[7]

> René Descartes walks into a restaurant and sits down for dinner. The waiter comes over and asks if he'd like an appetizer.
>
> "No thank you," says Descartes, "I'd just like to order dinner."
>
> "Would you like to hear our daily specials?" asks the waiter.
>
> "No," says Descartes, getting impatient.
>
> "Would you like a drink before dinner?" the waiter asks.
>
> Descartes is insulted, since he's a teetotaler. "I think not!" he says indignantly, and POOF! he disappears.
>
> —A joke as recalled by David Chalmers

There are two ways to view the questions we have been considering—converse Western and Eastern perspectives on the nature of consciousness and of reality. In the Western perspective, we start with a physical world that evolves patterns of information. After a few billion years of evolution, the entities in that world have evolved sufficiently to become conscious beings. In the Eastern view, consciousness is the fundamental reality; the physical world only comes into existence through the thoughts of conscious beings. The physical world, in other words, is the thoughts of conscious beings made manifest. These are of course simplifications of complex and diverse philosophies, but they represent the principal polarities in the philosophies of consciousness and its relationship to the physical world.

The East-West divide on the issue of consciousness has also found expression in opposing schools of thought in the field of subatomic physics. In quantum mechanics, particles exist as what are called probability fields. Any measurement carried out on them by a measuring device causes what is called a collapse of the wave function, meaning that the particle suddenly assumes a particular location. A popular view is that such a measurement constitutes observation by a conscious observer, because otherwise measurement would be a meaningless concept. Thus the particle assumes a particular location (as well as other properties, such as velocity) only when it is observed. Basically particles figure that if no one is bothering to look at them, they don't need to decide where they are. I call this the Buddhist school of quantum mechanics, because in it particles essentially don't exist until they are observed by a conscious person.

There is another interpretation of quantum mechanics that avoids such anthropomorphic terminology. In this analysis, the field representing a particle is not a probability field, but rather just a function that has different values in different locations. The field, therefore, is fundamentally what the particle is. There are constraints on what the values of the field can be in different locations, because the entire field representing a particle represents only a limited amount of information. That is where the word "quantum" comes from. The so-called collapse of the wave function, this view holds, is not a collapse at all. The wave function actually never goes away. It is just that a measurement device is also made up of particles with fields, and the interaction of the particle field being measured and the particle fields of the measuring device results in a reading of the particle being in a particular location. The field, however, is still present. This is the Western interpretation of quantum mechanics, although it is interesting to note that the more popular view among physicists worldwide is what I have called the Eastern interpretation.

There was one philosopher whose work spanned this East-West divide. The Austrian British thinker Ludwig Wittgenstein (1889–1951) studied the philosophy of language and knowledge and contemplated the question of what it is that we can really know. He pondered this subject while a soldier in World War I and took notes for what would be his only book published while he was alive, *Tractatus Logico-Philosophicus*. The work had an unusual structure, and it was only through the efforts of his former instructor, British mathematician and philosopher Bertrand Russell, that it found a publisher in 1921. It became the bible for a major school of philosophy known as logical positivism, which sought to define the limits of science. The book and the movement surrounding it were influential on Turing and the emergence of the theory of computation and linguistics.

Tractatus Logico-Philosophicus anticipates the insight that all knowledge is inherently hierarchical. The book itself is arranged in nested and numbered statements. For example, the first four statements in the book are:

> 1 The world is all that is the case.
> 1.1 The world is the totality of facts, not of things.
> 1.11 The world is determined by the facts, and by their being all the facts.
> 1.12 For the totality of facts determines what is the case, and also whatever is not the case.

Another significant statement in the *Tractatus*—and one that Turing would echo—is this:

> 4.0031 All philosophy is a critique of language.

Essentially both *Tractatus Logico-Philosophicus* and the logical positivism movement assert that physical reality exists separate from our perception of it, but that all we can know of that reality is what we perceive with our senses—which can be heightened through our tools—and the logical inferences we can make from these sensory impressions. Essentially Wittgenstein is attempting to describe the methods and goals of science. The final statement in the book is number 7, "What we cannot speak about we must pass over in silence." The early Wittgenstein, accordingly, considers the discussion of consciousness as circular and tautological and therefore a waste of time.

The later Wittgenstein, however, completely rejected this approach and spent all of his philosophical attention talking about matters that he had earlier argued should be passed over in silence. His writings on this revised thinking were collected and published in 1953, two years after his death, in a book called *Philosophical Investigations*. He criticized his earlier ideas in the *Tractatus,* judging them to be circular and void of meaning, and came to the view that what he had advised that we not speak about was in fact all that was worth reflecting on. These writings heavily influenced the existentialists, making Wittgenstein the only figure in modern philosophy to be a major architect of two leading and contradictory schools of thought in philosophy.

What is it that the later Wittgenstein thought was worth thinking and talking about? It was issues such as beauty and love, which he recognized exist imperfectly as ideas in the minds of men. However, he writes that such concepts do exist in a perfect and idealized realm, similar to the perfect "forms" that Plato wrote about in the Platonic dialogues, another work that illuminated apparently contradictory approaches to the nature of reality.

One thinker whose position I believe is mischaracterized is the French philosopher and mathematician René Descartes. His famous "I think, therefore I am" is generally interpreted to extol rational thought, in the sense that "I think, that is I can perform logical thought, therefore I am worthwhile." Descartes is therefore considered the architect of the Western rational perspective.

Reading this statement in the context of his other writings, however, I get a different impression. Descartes was troubled by what is referred to as the "mind-body problem": Namely, how does a conscious mind arise from the physical matter of the brain? From this perspective, it seems he was attempting to push rational skepticism to the breaking point, so in my view what his statement really means is, "I think, that is to say, a subjective experience is occurring, so therefore all we know for sure is that something—call it *I*—exists." He could not be certain that the physical world exists, because all we have are our own individual sense impressions of it, which might be wrong or completely illusory. We do know, however, that the experiencer exists.

My religious upbringing was in a Unitarian church, where we studied all of the world's religions. We would spend six months on, say, Buddhism and would go to Buddhist services, read their books, and have discussion groups with their leaders. Then we would switch to another religion, such as Judaism. The overriding theme was "many paths to the truth," along with tolerance and transcendence. This last idea meant that resolving apparent contradictions between traditions does not require deciding that one is right and the other is wrong. The truth can be discovered only by finding an explanation that overrides—transcends—seeming differences, especially for fundamental questions of meaning and purpose.

This is how I resolve the Western-Eastern divide on consciousness and the physical world. In my view, both perspectives have to be true.

On the one hand, it is foolish to deny the physical world. Even if we do live in a simulation, as speculated by Swedish philosopher Nick Bostrom, reality is nonetheless a conceptual level that is real for us. If we accept the existence of the physical world and the evolution that has taken place in it, then we can see that conscious entities have evolved from it.

On the other hand, the Eastern perspective—that consciousness is fundamental and represents the only reality that is truly important—is also difficult to deny. Just consider the precious regard we give to conscious persons versus unconscious things. We consider the latter to have no intrinsic value except to the extent that they can influence the subjective experience of conscious persons. Even if we regard consciousness as an emergent property of a complex system, we cannot take the position that it is just another attribute (along with "digestion" and "lactation," to quote John Searle). It represents what is truly important.

The word "spiritual" is often used to denote the things that are of ultimate significance. Many people don't like to use such terminology from spiritual or religious traditions, because it implies sets of beliefs that they may not subscribe to. But if we strip away the mystical complexities of religious traditions and simply respect "spiritual" as implying something of profound meaning to humans, then the concept of consciousness fits the bill. It reflects the ultimate spiritual value. Indeed, "spirit" itself is often used to denote consciousness.

Evolution can then be viewed as a spiritual process in that it creates spiritual beings, that is, entities that are conscious. Evolution also moves toward greater complexity, greater knowledge, greater intelligence, greater beauty, greater creativity, and the ability to express more transcendent emotions, such as love. These are all descriptions that people have used for the concept of God, albeit God is described as having no limitations in these regards.

People often feel threatened by discussions that imply the possibility that a machine could be conscious, as they view considerations along these lines as a denigration of the spiritual value of conscious persons. But this reaction reflects a misunderstanding of the concept of a machine. Such critics are addressing the issue based on the machines they know today, and as impressive as they are becoming, I agree that contemporary examples of technology are not yet worthy of our respect as conscious beings. My prediction is that they will become indistinguishable from biological humans, whom we do regard as conscious beings, and will therefore share in the spiritual value we ascribe to consciousness. This is not a disparagement of people; rather, it is an elevation of our understanding of (some) future machines. We should probably adopt a different terminology for these entities, as they will be a different sort of machine.

Indeed, as we now look inside the brain and decode its mechanisms we discover methods and algorithms that we can not only understand but recreate—"the parts of a mill pushing on each other," to paraphrase German mathematician and philosopher Gottfried Wilhelm Leibniz (1646–1716) when he wrote about the brain. Humans already constitute spiritual machines. Moreover, we will merge with the tools we are creating so closely that the distinction between human and machine will blur until the difference disappears. That process is already well under way, even if most of the machines that extend us are not yet inside our bodies and brains.

Free Will

A central aspect of consciousness is the ability to look ahead, the capability we call "foresight." It is the ability to plan, and in social terms to outline a scenario of what is likely going to happen, or what might happen, in social interactions that have not yet taken place. . . . It is a system whereby we improve our chances of doing those things that will represent our own best interests. . . . I suggest that "free will" is our apparent ability to choose and act upon whichever of those seem most useful or appropriate, and our insistence upon the idea that such choices are our own.

—Richard D. Alexander

Shall we say that the plant does not know what it is doing merely because it has no eyes, or ears, or brains? If we say that it acts mechanically, and mechanically only, shall we not be forced to admit that sundry other and apparently very deliberate actions are also mechanical? If it seems to us that the plant kills and eats a fly mechanically, may it not seem to the plant that a man must kill and eat a sheep mechanically?

—Samuel Butler, 1871

Is the brain, which is notably double in structure, a double organ, "seeming parted, but yet a union in partition"?

—Henry Maudsley[8]

Redundancy, as we have learned, is a key strategy deployed by the neocortex. But there is another level of redundancy in the brain, in that its left and right hemispheres, while not identical, are largely the same. Just as certain regions of the neocortex normally end up processing certain types of information, the hemispheres also specialize to some extent—for example, the left hemisphere typically is responsible for verbal language. But these assignments can also be rerouted, to the point that we can survive and function somewhat normally with only one half. American neuropsychology researchers Stella de Bode and Susan Curtiss reported on forty-nine children who had undergone a hemispherectomy (removal of half of their brain), an extreme operation that is performed on patients with a life-threatening seizure disorder that exists in only one hemisphere. Some who undergo the procedure are left with deficits, but those deficits are specific and the patients have reasonably normal personalities. Many of them thrive, and it is not apparent to observers that they only have half a brain. De Bode and Curtiss write about left-hemispherectomized children who "develop remarkably good language despite removal of the 'language' hemisphere."[9] They describe one such student who completed college, attended graduate school, and scored above average on IQ tests. Studies have shown minimal long-term effects on overall cognition, memory, personality, and sense of humor.[10] In a 2007 study American researchers Shearwood McClelland and Robert Maxwell showed similar long-term positive results in adults.[11]

A ten-year-old German girl who was born with only half of her brain has also been reported to be quite normal. She even has almost perfect vision in one eye, whereas hemispherectomy patients lose part of their field of vision right after the operation.[12] Scottish researcher Lars Muckli commented, "The brain has amazing plasticity but we were quite astonished to see just how well the single hemisphere of the brain in this girl has adapted to compensate for the missing half."

While these observations certainly support the idea of plasticity in the neocortex, their more interesting implication is that we each appear to have two brains, not one, and we can do pretty well with either. If we lose one, we do lose the cortical patterns that are uniquely stored there, but each brain is in itself fairly complete. So does each hemisphere have its own consciousness? There is an argument to be made that such is the case.

Consider split-brain patients, who still have both of their brain hemispheres, but the channel between them has been cut. The corpus callosum is a bundle of about 250 million axons that connects the left and right cerebral hemispheres and enables them to communicate and coordinate with each other. Just as two people can communicate closely with each other and act as a single decision maker while remaining separate and whole individuals, the two brain hemispheres can function as a unit while remaining independent.

As the term implies, in split-brain patients the corpus callosum has been cut or damaged, leaving them effectively with two functional brains without a direct communication link between them. American psychology researcher Michael Gazzaniga (born in 1939) has conducted extensive experiments on what each hemisphere in split-brain patients is thinking.

The left hemisphere in a split-brain patient usually sees the right visual field, and vice versa. Gazzaniga and his colleagues showed a split-brain patient a picture of a chicken claw to the right visual field (which was seen by his left hemisphere) and a snowy scene to the left visual field (which was seen by his right hemisphere). He then showed a collection of pictures so that both hemispheres could see them. He asked the patient to choose one of the pictures that went well with the first picture. The patient's left hand (controlled by his right hemisphere) pointed to a picture of a shovel, whereas his right hand pointed to a picture of a chicken. So far so good— the two hemispheres were acting independently and sensibly. "Why did you choose that?" Gazzaniga asked the patient, who answered verbally (controlled by his left-hemisphere speech center), "The chicken claw obviously goes with the chicken." But then the patient looked down and, noticing his left hand pointing to the shovel, immediately explained this (again with his left-hemisphere-controlled speech center) as "and you need a shovel to clean out the chicken shed."

This is a confabulation. The right hemisphere (which controls the left arm and hand) correctly points to the shovel, but because the left hemisphere (which controls the verbal answer) is unaware of the snow, it confabulates an explanation, yet is not aware that it is confabulating. It is taking responsibility for an action it had never decided on and never took, but thinks that it did.

This implies that each of the two hemispheres in a split-brain patient has its own consciousness. The hemispheres appear not to be aware that their body is effectively controlled by two brains, because they learn to coordinate with each other, and their decisions are sufficiently aligned and consistent that each thinks that the decisions of the other are its own.

Gazzaniga's experiment doesn't prove that a normal individual with a functioning corpus callosum has two conscious half-brains, but it is suggestive of that possibility. While the corpus callosum allows for effective collaboration between the two half-brains, it doesn't necessarily mean that they are not separate minds. Each one could be fooled into thinking it has made all the decisions, because they would all be close enough to what each would have decided on its own, and after all, it does have a lot of influence on each decision (by collaborating with the other hemisphere through the corpus callosum). So to each of the two minds it would seem as if it were in control.

How would you test the conjecture that they are both conscious? One could assess them for neurological correlates of consciousness, which is precisely what Gazzaniga has done. His experiments show that each hemisphere is acting as an independent brain. Confabulation is not restricted to brain hemispheres; we each do it on a regular basis. Each hemisphere is about as intelligent as a human, so if we believe that a human brain is conscious, then we have to conclude that each hemisphere is independently conscious. We can assess the neurological correlates and we can conduct our own thought experiments (for example, considering that if two brain hemispheres without a functioning corpus callosum constitute two separate conscious minds, then the same would have to hold true for two hemispheres with a functioning connection between them), but any attempt at a more direct detection of consciousness in each hemisphere confronts us again with the lack of a scientific test for consciousness. But if we do allow that each hemisphere of the brain is conscious, then do we grant that the so-called unconscious activity in the neocortex (which constitutes the vast bulk of its activity) has an independent consciousness too? Or maybe it has more than one? Indeed, Marvin Minsky refers to the brain as a "society of mind."[13]

In another split-brain experiment the researchers showed the word "bell" to the right brain and "music" to the left brain. The patient was asked what word he saw. The left-hemisphere-controlled speech center says "music." The subject was then shown a group of pictures and asked to point to a picture most closely related to the word he was just shown. His right-hemisphere-controlled arm pointed to the bell. When he was asked why he pointed to the bell, his left-hemisphere-controlled speech center replied, "Well, music, the last time I heard any music was the bells banging outside here." He provided this explanation even though there were other pictures to choose from that were much more closely related to music.

Again, this is a confabulation. The left hemisphere is explaining as if it were its own a decision that it never made and never carried out. It is not doing so to cover up for a friend (that is, its other hemisphere)—it genuinely thinks that the decision was its own.

These reactions and decisions can extend to emotional responses. They asked a teenage split-brain patient—so that both hemispheres heard—"Who is your favorite . . ." and then fed the word "girlfriend" just to the right hemisphere through the left ear. Gazzaniga reports that the subject blushed and acted embarrassed, an appropriate reaction for a teenager when asked about his girlfriend. But the left-hemisphere-controlled speech center reported that it had not heard any word and asked for clarification: "My favorite what?" When asked again to answer the question, this time in writing, the right-hemisphere-controlled left hand wrote out his girlfriend's name.

Gazzaniga's tests are not thought experiments but actual mind experiments. While they offer an interesting perspective on the issue of consciousness, they speak even more directly to the issue of free will. In each of these cases, one of the hemispheres believes that it has made a decision that it in fact never made. To what extent is that true for the decisions we make every day?

Consider the case of a ten-year-old female epileptic patient. Neurosurgeon Itzhak Fried was performing brain surgery while she was awake (which is feasible because there are no pain receptors in the brain).[14] Whenever he stimulated a particular spot on her neocortex, she would laugh. At first the surgical team thought that they might be triggering some sort of laugh reflex, but they quickly realized that they were triggering the actual perception of humor. They had apparently found a point in her neocortex—there is obviously more than one—that recognizes the perception of humor. She was not just laughing—she actually found the situation funny, even though nothing had actually changed in the situation other than their having stimulated this point in her neocortex. When they asked her why she was laughing, she did not reply along the lines of, "Oh, no particular reason," or "You just stimulated my brain," but would immediately confabulate a reason. She would point to something in the room and try to explain why it was funny. "You guys are just so funny standing there" was a typical comment.

We are apparently very eager to explain and rationalize our actions, even when we didn't actually make the decisions that led to them. So just how responsible are we for our decisions? Consider these experiments by physiology professor Benjamin Libet (1916–2007) at the University of California at Davis. Libet had participants sit in front of a timer, EEG electrodes attached to their scalps. He instructed them to do simple tasks such as pushing a button or moving a finger. The participants were asked to note the time on the timer when they "first become aware of the wish or urge to act." Tests indicated a margin of error of only 50 milliseconds on these assessments by the subjects. They also measured an average of about 200 milliseconds between the time when the subjects reported awareness of the urge to act and the actual act.[15]

The researchers also looked at the EEG signals coming from the subjects' brains. Brain activity involved in initiating the action by the motor cortex (which is responsible for carrying out the action) actually occurred on average about 500 milliseconds prior to the performance of the task. That means that the motor cortex was preparing to carry out the task about a third of a second before the subject was even aware that she had made a decision to do so.

The implications of the Libet experiments have been hotly debated. Libet himself concluded that our awareness of decision making appears to be an illusion, that "consciousness is out of the loop." Philosopher Daniel Dennett commented, "The action is originally precipitated in some part of the brain, and off fly the signals to muscles, pausing en route to tell you, the conscious agent, what is going on (but like all good officials letting you, the bumbling president, maintain the illusion that you started it all)."[16] At the same time Dennett has questioned the timings recorded by the experiment, basically arguing that subjects may not really be aware of when they become aware of the decision to act. One might wonder: If the subject is unaware of when she is aware of making a decision, then who is? But the point is actually well taken—as I discussed earlier, what we are conscious of is far from clear.

Indian American neuroscientist Vilayanur Subramanian "Rama" Ramachandran (born in 1951) explains the situation a little differently. Given that we have on the order of 30 billion neurons in the neocortex, there is always a lot going on there, and we are consciously aware of very little of it. Decisions, big and little, are constantly being processed by the neocortex, and proposed solutions bubble up to our conscious awareness. Rather than free will, Ramachandran suggests we should talk about "free won't"—that is, the power to reject solutions proposed by the nonconscious parts of our neocortex.

Consider the analogy to a military campaign. Army officials prepare a recommendation to the president. Prior to receiving the president's approval, they perform preparatory work that will enable the decision to be carried out. At a particular moment, the proposed decision is presented to the president, who approves it, and the rest of the mission is then undertaken. Since the "brain" represented by this analogy involves the unconscious processes of the neocortex (that is, the officials under the president) as well as its conscious processes (the president), we would see neural activity as well as actual actions taking place prior to the official decision's being made. We can always get into debates in a particular situation as to how much leeway the officials under the president actually gave him or her to accept or reject a recommendation, and certainly American presidents have done both. But it should not surprise us that mental activity, even in the motor cortex, would start before we were aware that there was a decision to be made.

What the Libet experiments do underscore is that there is a lot of activity in our brains underlying our decisions that is not conscious. We already knew that most of what goes in the neocortex is not conscious; it should not be surprising, therefore, that our actions and decisions stem from both unconscious and conscious activity. Is this distinction important? If our decisions arise from both, should it matter if we sort out the conscious parts from the unconscious? Is it not the case that both aspects represent our brain? Are we not ultimately responsible for everything that goes on in our brains? "Yes, I shot the victim, but I'm not responsible because I wasn't paying attention" is probably a weak defense. Even though there are some narrow legal grounds on which a person may not be held responsible for his decisions, we are generally held accountable for all of the choices we make.

The observations and experiments I have cited above constitute thought experiments on the issue of free will, a subject that, like the topic of consciousness, has been debated since Plato. The term "free will" itself dates back to the thirteenth century, but what exactly does it mean?

The Merriam-Webster dictionary defines it as the "freedom of humans to make choices that are not determined by prior causes or by divine intervention." You will notice that this definition is hopelessly circular: "Free will is freedom. . . ." Setting aside the idea of divine intervention's standing in opposition to free will, there is one useful element in this definition, which is the idea of a decision's "not [being] determined by prior causes." I'll come back to that momentarily.

The Stanford Encyclopedia of Philosophy states that free will is the "capacity of rational agents to choose a course of action from among various alternatives." By this definition, a simple computer is capable of free will, so it is less helpful than the dictionary definition.

Wikipedia is actually a bit better. It defines free will as "the ability of agents to make choices free from certain kinds of constraints. . . . The constraint of dominant concern has been . . . determinism." Again, it uses the circular word "free" in defining free will, but it does articulate what has been regarded as the principal enemy of free will: *determinism*. In that respect the Merriam-Webster definition above is actually similar in its reference to decisions that "are not determined by prior causes."

So what do we mean by determinism? If I put "2 + 2" into a calculator and it displays "4," can I say that the calculator displayed its free will by deciding to display that "4"? No one would accept that as a demonstration of free will, because the "decision" was predetermined by the internal mechanisms of the calculator and the input. If I put in a more complex calculation, we still come to the same conclusion with regard to its lack of free will.

How about Watson when it answers a *Jeopardy!* query? Although its deliberations are far more complex than those of the calculator, very few if any observers would ascribe free will to its decisions. No one human knows exactly how all of its programs work, but we can identify a group of people who collectively can describe all of its methods. More important, its output is determined by (1) all of its programs at the moment that the query is posed, (2) the query itself, (3) the state of its internal parameters that influence its decisions, and (4) its trillions of bytes of knowledge bases, including encyclopedias. Based on these four categories of information, its output is determined. We might speculate that presenting the same query would always get the same response, but Watson is programmed to learn from its experience, so there is the possibility that subsequent answers would be different. However, that does not contradict this analysis; rather, it just constitutes a change in item 3, the parameters that control its decisions.

So how exactly does a human differ from Watson, such that we ascribe free will to the human but not to the computer program? We can identify several factors. Even though Watson is a better *Jeopardy!* player than most if not all humans, it is nonetheless not nearly as complex as a human neo-cortex. Watson does possess a lot of knowledge, and it does use hierar-chical methods, but the complexity of its hierarchical thinking is still considerably less than that of a human. So is the difference simply one of the scale of complexity of its hierarchical thinking? There is an argument to be made that the issue does come down to this. In my discussion of the issue of consciousness I noted that my own leap of faith is that I would consider a computer that passed a valid Turing test to be conscious. The best chatbots are not able to do that today (although they are steadily improving), so my conclusion with regard to consciousness is a matter of the level of performance of the entity. Perhaps the same is true of my ascribing free will to it.

Consciousness is indeed one philosophical difference between human brains and contemporary software programs. We consider human brains to be conscious, whereas we do not—*yet*—attribute that to software pro-grams. Is this the factor we are looking for that underlies free will?

A simple mind experiment would argue that consciousness is indeed a vital part of free will. Consider a situation in which someone performs an action with no awareness that she is doing it—it is carried out entirely by nonconscious activity in that person's brain. Would we regard this to be a display of free will? Most people would answer no. If the action was harm-ful, we would probably still hold that person responsible but look for some recent conscious acts that may have caused that person to perform actions without conscious awareness, such as taking one drink too many, or just failing to train herself adequately to consciously consider her decisions before she acted on them.

According to some commentators, the Libet experiments argued against free will by highlighting how much of our decision making is not conscious. Since there is a reasonable consensus among philosophers that free will does imply conscious decision making, it appears to be one prerequisite for free will. However, to many observers, consciousness is a necessary but not sufficient condition. If our decisions—conscious or otherwise—are predetermined before we make them, how can we say that our decisions are free? This position, which holds that free will and determinism are not compatible, is known as incompatibilism. For example, American philosopher Carl Ginet (born in 1932) argues that if events in the past, present, and future are determined, then we can be considered to have no control over them or their consequences. Our apparent decisions and actions are simply part of this predetermined sequence. To Ginet, this rules out free will.

Not everyone regards determinism as being incompatible with the concept of free will, however. The compatibilists argue, essentially, that you're free to decide what you want even though what you decide is or may be determined. Daniel Dennett, for example, argues that while the future may be determined from the state of the present, the reality is that the world is so intricately complex that we cannot possibly know what the future will bring. We can identify what he refers to as "expectations," and we are indeed free to perform acts that differ from these expectations. We should consider how our decisions and actions compare to these expectations, not to a theoretically determined future that we cannot in fact know. That, Dennett argues, is sufficient for free will.

Gazzaniga also articulates a compatibilist position: "We are personally responsible agents and are to be held accountable for our actions, even though we live in a determined world."[17] A cynic might interpret this view as: You have no control over your actions, but we'll blame you anyway.

Some thinkers dismiss the idea of free will as an illusion. Scottish philosopher David Hume (1711–1776) described it as simply a "verbal" matter characterized by "a false sensation or seeming experience."[18] German philosopher Arthur Schopenhauer (1788–1860) wrote that "everyone believes himself *a priori* to be perfectly free, even in his individual actions, and thinks that at every moment he can commence another manner of life. . . . But *a posteriori,* through experience, he finds to his astonishment that he is not free, but subjected to necessity, that in spite of all his resolutions and reflections he does not change his conduct, and that from the beginning of his life to the end of it, he must carry out the very character which he himself condemns."[19]

I would add several points here. The concept of free will—and responsibility, which is a closely aligned idea—is useful, and indeed vital, to maintaining social order, whether or not free will actually exists. Just as consciousness clearly exists as a meme, so too does free will. Attempts to prove its existence, or even to define it, may become hopelessly circular, but the reality is that almost everyone believes in the idea. Very substantial portions of our higher-level neocortex are devoted to the concept that we make free choices and are responsible for our actions. Whether in a strict philosophical sense that is true or even possible, society would be far worse off if we did not have such beliefs.

Furthermore, the world is not necessarily determined. I discussed above two perspectives on quantum mechanics, which differ with respect to the relationship of quantum fields to an observer. A popular interpretation of the observer-based perspective provides a role for consciousness: Particles do not resolve their quantum ambiguity until observed by a conscious observer. There is another split in the philosophy of quantum events that has a bearing on our discussion of free will, one that revolves around the question: Are quantum events determined or random?

The most common interpretation of a quantum event is that when the wave function constituting a particle "collapses," the particle's location becomes specific. Over a great many such events, there will be a predictable distribution (which is why the wave function is considered to be a probability distribution), but the resolution for each such particle undergoing a collapse of its wave function is random. The opposing interpretation is deterministic: specifically, that there is a hidden variable that we are unable to detect separately, but whose value determines the particle's position. The value or phase of the hidden variable at the moment of the wave function collapse determines the position of the particle. Most quantum physicists seem to favor the idea of a random resolution according to the probability field, but the equations for quantum mechanics do allow for the existence of such a hidden variable.

Thus the world may not be determined after all. According to the probability wave interpretation of quantum mechanics, there is a continual source of uncertainty at the most basic level of reality. However, this observation does not necessarily resolve the concerns of the incompatibilists. It is true that under this interpretation of quantum mechanics, the world is not determined, but our concept of free will extends beyond decisions and actions that are merely random. Most incompatibilists would find the concept of free will to also be incompatible with our decisions' being essentially accidental. Free will seems to imply purposeful decision making.

Dr. Wolfram proposes a way to resolve the dilemma. His book *A New Kind of Science* (2002) presents a comprehensive view of the idea of cellular automata and their role in every facet of our lives. A cellular automaton is a mechanism in which the value of information cells is continually recomputed as a function of the cells near it. John von Neumann created a theoretical self-replicating machine called a universal constructor that was perhaps the first cellular automaton.

Dr. Wolfram illustrates his thesis with the simplest possible cellular automata, a group of cells in a one-dimensional line. At each point in time, each cell can have one of two values: black or white. The value of each cell is recomputed for each cycle. The value of a cell for the next cycle is a function of its current value as well as the value of its two adjacent neighbors. Each cellular automaton is characterized by a rule that determines how we compute whether a cell is black or white in the next cycle.

Consider the example of what Dr. Wolfram calls rule 222.

rule 222

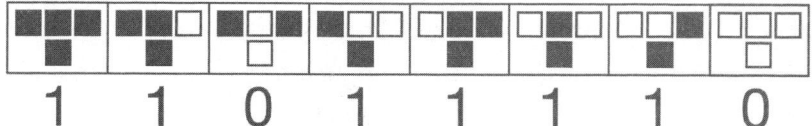

The eight possible combinations of value for the cell being recomputed and its left and right neighbors are shown in the top row. Its new value is shown in the bottom row. So, for example, if the cell is black and its two neighbors are also black, then the cell will remain black in the next generation (see the leftmost subrule of rule 222). If the cell is white, its left neighbor is white, and its right neighbor is black, then it will be changed to black in the next generation (see the subrule of rule 222 that is second from the right).

The universe for this simple cellular automaton is just one row of cells. If we start with just one black cell in the middle and show the evolution of the cells over multiple generations (where each row as we move down represents a new generation of values), the results of rule 222 look like this:

rule 222

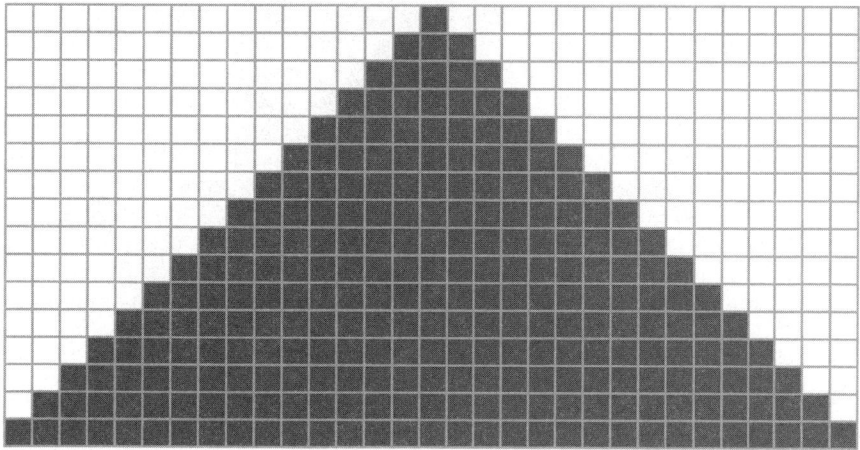

An automaton is based on a rule, and a rule defines whether the cell will be black or white based on which of the eight possible patterns exist in the current generation. Thus there are $2^8 = 256$ possible rules. Dr. Wolfram listed all 256 possible such automata and assigned each a Wolfram code from 0 to 255. Interestingly, these 256 theoretical machines have very different properties. The automata in what Dr. Wolfram calls class I, such as rule 222, create very predictable patterns. If I were to ask what the value of the middle cell was after a trillion trillion iterations of rule 222, you could answer easily: black.

Much more interesting, however, are the class IV automata, illustrated by rule 110.

rule 110

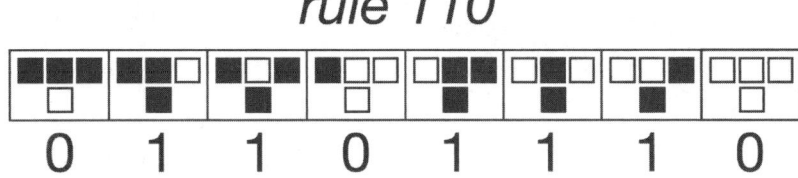

Multiple generations of this automaton look like this:

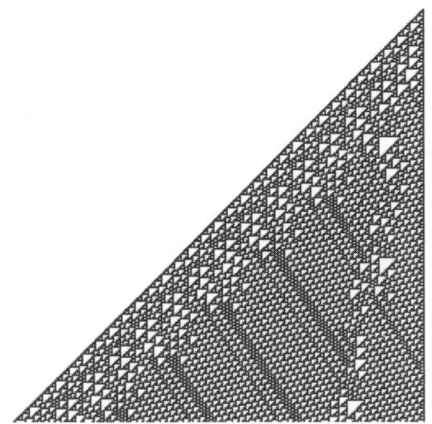

The interesting thing about the rule 110 automaton and class IV automata in general is that the results are completely unpredictable. The results pass the strictest mathematical tests for randomness, yet they do not simply generate noise: There are repeating patterns, but they repeat in odd and unpredictable ways. If I were to ask you what the value of a particular cell was after a trillion trillion iterations, there would be no way to answer that question without actually running this machine through that many generations. The solution is clearly determined, because this is a very simple deterministic machine, but it is completely unpredictable without actually running the machine.

Dr. Wolfram's primary thesis is that the world is one big class IV cellular automaton. The reason that his book is titled *A New Kind of Science* is because this theory contrasts with most other scientific laws. If there is a satellite orbiting Earth, we can predict where it will be five years from now without having to run through each moment of a simulated process by using the relevant laws of gravity and solve where it will be at points in time far in the future. But the future state of class IV cellular automata cannot be predicted without simulating every step along the way. If the universe is a giant cellular automaton, as Dr. Wolfram postulates, there would be no computer big enough—since every computer would be a subset of the universe—that could run such a simulation. Therefore the future state of the universe is completely unknowable even though it is deterministic.

Thus even though our decisions are determined (because our bodies and brains are part of a deterministic universe), they are nonetheless inherently unpredictable because we live in (and are part of) a class IV automaton. We cannot predict the future of a class IV automaton except to let the future unfold. For Dr. Wolfram, this is sufficient to allow for free will.

We don't have to look to the universe to see future events that are determined yet unpredictable. None of the scientists who have worked on Watson can predict what it will do, because the program is just too complex and varied, and its performance is based on knowledge that is far too extensive for any human to master. If we believe that humans exhibit free will, then it follows that we have to allow that future versions of Watson or Watson-like machines can exhibit it also.

My own leap of faith is that I believe that humans have free will, and while I act as if that is the case, I am hard pressed to find examples among my own decisions that illustrate that. Consider the decision to write this book—I never made that decision. Rather, the idea of the book decided that for me. In general, I find myself captive to ideas that seem to implant themselves in my neocortex and take over. How about the decision to get married, which I made (in collaboration with one other person) thirty-six years ago? At the time, I had been following the usual program of being attracted to—and pursuing—a pretty girl. I then fell in love. Where is the free will in that?

But what about the little decisions I make every day—for example, the specific words I choose to write in my book? I start with a blank virtual sheet of paper. No one is telling me what to do. There is no editor looking over my shoulder. My choices are *entirely* up to me. I am free—*totally free*—to write *whatever* I . . .

Uh, *grok* . . .

Grok? Okay, I did it—I finally applied my free will. I was going to write the word "want," but I made a free decision to write something totally unexpected instead. This is perhaps the first time I've succeeded in exercising pure free will.

Or not.

It should be apparent that that was a display not of will, but rather of trying to illustrate a point (and perhaps a weak sense of humor).

Although I share Descartes' confidence that I am conscious, I'm not so sure about free will. It is difficult to escape Schopenhauer's conclusion that "you can do what you will, but in any given moment of your life you can *will* only one definite thing and absolutely nothing other than that one thing."[20] Nonetheless I will continue to act as if I have free will and to believe in it, so long as I don't have to explain why.

Identity

A philosopher once had the following dream.

First Aristotle appeared, and the philosopher said to him, "Could you give me a fifteen-minute capsule sketch of your entire philosophy?"

To the philosopher's surprise, Aristotle gave him an excellent exposition in which he compressed an enormous amount of material into a mere fifteen minutes. But then the philosopher raised a certain objection which Aristotle couldn't answer. Confounded, Aristotle disappeared.

Then Plato appeared. The same thing happened again, and the philosopher's objection to Plato was the same as his objection to Aristotle. Plato also couldn't answer it and disappeared.

Then all the famous philosophers of history appeared one by one and our philosopher refuted every one with the same objection.

After the last philosopher vanished, our philosopher said to himself, "I know I'm asleep and dreaming all this. Yet I've found a universal refutation for all philosophical systems! Tomorrow when I wake up, I will probably have forgotten it, and the world will really miss something!" With an iron effort, the philosopher forced himself to wake up, rush over to his desk, and write down his universal refutation. Then he jumped back into bed with a sigh of relief.

The next morning when he awoke, he went over to the desk to see what he had written. It was, "That's what *you* say."

—Raymond Smullyan, as quoted by David Chalmers[21]

What I wonder about ever more than whether or not I am conscious or exercise free will is why I happen to be conscious of the experiences and decisions of this one particular person who writes books, enjoys hiking and biking, takes nutritional supplements, and so on. An obvious answer would be, "Because that's who you are."

That exchange is probably no more tautological than my answers above to questions about consciousness and free will. But actually I do have a better answer for why my consciousness is associated with this particular person: It is because that is who I created myself to be.

A common aphorism is, "You are what you eat." It is even more true to say, "You are what you think." As we have discussed, all of the hierarchical structures in my neocortex that define my personality, skills, and knowledge are the result of my own thoughts and experiences. The people I choose to interact with and the ideas and projects I choose to engage in are all primary determinants of who I become. For that matter, what I eat also reflects the decisions made by my neocortex. Accepting the positive side of the free will duality for the moment, it is my own decisions that result in who I am.

Regardless of how we came to be who we are, each of us has the desire for our identity to persist. If you didn't have the will to survive, you wouldn't be here reading this book. Every creature has that goal—it is the principal determinant of evolution. The issue of identity is perhaps even harder to define than consciousness or free will, but is arguably more important. After all, we need to know what we are if we seek to preserve our existence.

Consider this thought experiment: You are in the future with technologies more advanced than today's. While you are sleeping, some group scans your brain and picks up every salient detail. Perhaps they do this with blood cell–sized scanning machines traveling in the capillaries of your brain or with some other suitable noninvasive technology, but they have all of the information about your brain at a particular point in time. They also pick up and record any bodily details that might reflect on your state of mind, such as the endocrine system. They instantiate this "mind file" in a nonbiological body that looks and moves like you and has the requisite subtlety and suppleness to pass for you. In the morning you are informed about this transfer and you watch (perhaps without being noticed) your mind clone, whom we'll call You 2. You 2 is talking about his or her life as if s/he were you, and relating how s/he discovered that very morning that s/he had been given a much more durable new version 2.0 body. "Hey, I kind of like this new body!" s/he exclaims.

The first question to consider is: Is You 2 conscious? Well, s/he certainly seems to be. S/he passes the test I articulated earlier, in that s/he has the subtle cues of being a feeling, conscious person. If you are conscious, then so too is You 2.

So if you were to, uh, disappear, no one would notice. You 2 would go around claiming to be you. All of your friends and loved ones would be content with the situation and perhaps pleased that you now have a more durable body and mental substrate than you used to have. Perhaps your more philosophically minded friends would express concerns, but for the most part, everybody would be happy, including you, or at least the person who is convincingly claiming to be you.

So we don't need your old body and brain anymore, right? Okay if we dispose of it?

You're probably not going to go along with this. I indicated that the scan was noninvasive, so you are still around and still conscious. Moreover your sense of identity is still with you, not with You 2, even though You 2 thinks s/he is a continuation of you. You 2 might not even be aware that you exist or ever existed. In fact you would not be aware of the existence of You 2 either, if we hadn't told you about it.

Our conclusion? You 2 is conscious but is a different person than you—You 2 has a different identity. S/he is extremely similar, much more so than a mere genetic clone, because s/he also shares all of your neocortical patterns and connections. Or I should say s/he shared those patterns at the moment s/he was created. At that point, the two of you started to go your own ways, neocortically speaking. You are still around. You are not having the same experiences as You 2. Bottom line: You 2 is not you.

Okay, so far so good. Now consider another thought experiment—one that is, I believe, more realistic in terms of what the future will bring. You undergo a procedure to replace a very small part of your brain with a non-biological unit. You're convinced that it's safe, and there are reports of various benefits.

This is not so far-fetched, as it is done routinely for people with neurological and sensory impairments, such as the neural implant for Parkinson's disease and cochlear implants for the deaf. In these cases the computerized device is placed inside the body but outside the brain yet connected into the brain (or in the case of the cochlear implants, to the auditory nerve). In my view the fact that the actual computer is physically placed outside the actual brain is not philosophically significant: We are effectively augmenting the brain and replacing with a computerized device those of its functions that no longer work properly. In the 2030s, when intelligent computerized devices will be the size of blood cells (and keep in mind that white blood cells are sufficiently intelligent to recognize and combat pathogens), we will introduce them noninvasively, no surgery required.

Returning to our future scenario, you have the procedure, and as promised, it works just fine—certain of your capabilities have improved. (You have better memory, perhaps.) So are you still you? Your friends certainly think so. You think so. There is no good argument that you're suddenly a different person. Obviously, you underwent the procedure in order to effect a change in something, but you are still the same you. Your identity hasn't changed. Someone else's consciousness didn't suddenly take over your body.

Okay, so, encouraged by these results, you now decide to have another procedure, this time involving a different region of the brain. The result is the same: You experience some improvement in capability, but you're still you.

It should be apparent where I am going with this. You keep opting for additional procedures, your confidence in the process only increasing, until eventually you've changed every part of your brain. Each time the procedure was carefully done to preserve all of your neocortical patterns and connections so that you have not lost any of your personality, skills, or memories. There was never a you and a You 2; there was only you. No one, including you, ever notices you ceasing to exist. Indeed—there you are.

Our conclusion: You still exist. There's no dilemma here. Everything is fine.

Except for this: You, after the gradual replacement process, are entirely equivalent to You 2 in the prior thought experiment (which I will call the scan-and-instantiate scenario). You, after the gradual replacement scenario, have all of the neocortical patterns and connections that you had originally, only in a nonbiological substrate, which is also true of You 2 in the scan-and-instantiate scenario. You, after the gradual replacement scenario, have some additional capabilities and greater durability than you did before the process, but this is likewise true of You 2 in the scan-and-instantiate process.

But we concluded that You 2 is *not* you. And if you, after the gradual replacement process, are entirely equivalent to You 2 after the scan-and-instantiate process, then you after the gradual replacement process must also not be you.

That, however, contradicts our earlier conclusion. The gradual replacement process consists of multiple steps. Each of those steps appeared to preserve identity, just as we conclude today that a Parkinson's patient has the same identity after having had a neural implant installed.[22]

It is just this sort of philosophical dilemma that leads some people to conclude that these replacement scenarios will never happen (even though they are already taking place). But consider this: We naturally undergo a gradual replacement process throughout our lives. Most of our cells in our body are continuously being replaced. (You just replaced 100 million of them in the course of reading the last sentence.) Cells in the inner lining of the small intestine turn over in about a week, as does the stomach's protective lining. The life span of white blood cells ranges from a few days to a few months, depending on the type. Platelets last about nine days.

Neurons persist, but their organelles and their constituent molecules turn over within a month.[23] The half-life of a neuron microtubule is about ten minutes; the actin filaments in the dendrites last about forty seconds; the proteins that provide energy to the synapses are replaced every hour; the NMDA receptors in synapses are relatively long-lived at five days.

So you are completely replaced in a matter of months, which is comparable to the gradual replacement scenario I describe above. Are you the same person you were a few months ago? Certainly there are some differences. Perhaps you learned a few things. But you assume that your identity persists, that you are not continually destroyed and re-created.

Consider a river, like the one that flows past my office. As I look out now at what people call the Charles River, is it the same river that I saw yesterday? Let's first reflect on what a river is. The dictionary defines it is "a large natural stream of flowing water." By that definition, the river I'm looking at is a completely different one than it was yesterday. Every one of its water molecules has changed, a process that happens very quickly. Greek philosopher Diogenes Laertius wrote in the third century AD that "you cannot step into the same river twice."

But that is not how we generally regard rivers. People like to look at them because they are symbols of continuity and stability. By the common view, the Charles River that I looked at yesterday is the same river I see today. Our lives are much the same. Fundamentally we are not the stuff that makes up our bodies and brains. These particles essentially flow through us in the same way that water molecules flow through a river. We are a pattern that changes slowly but has stability and continuity, even though the stuff constituting the pattern changes quickly.

The gradual introduction of nonbiological systems into our bodies and brains will be just another example of the continual turnover of parts that compose us. It will not alter the continuity of our identity any more than the natural replacement of our biological cells does. We have already largely outsourced our historical, intellectual, social, and personal memories to our devices and the cloud. The devices we interact with to access these memories may not yet be inside our bodies and brains, but as they become smaller and smaller (and we are shrinking technology at a rate of about a hundred in 3-D volume per decade), they will make their way there. In any event, it will be a useful place to put them—we won't lose them that way. If people do opt out of placing microscopic devices inside their bodies, that will be fine, as there will be other ways to access the pervasive cloud intelligence.

But we come back to the dilemma I introduced earlier. You, after a period of gradual replacement, are equivalent to You 2 in the scan-

and-instantiate scenario, but we decided that You 2 in that scenario does not have the same identity as you. So where does that leave us?

It leaves us with an appreciation of a capability that nonbiological systems have that biological systems do not: the ability to be copied, backed up, and re-created. We do that routinely with our devices. When we get a new smartphone, we copy over all of our files, so it has much the same personality, skills, and memories that the old smartphone did. Perhaps it also has some new capabilities, but the contents of the old phone are still with us. Similarly, a program such as Watson is certainly backed up. If the Watson hardware were destroyed tomorrow, Watson would easily be re-created from its backup files stored in the cloud.

This represents a capability in the nonbiological world that does not exist in the biological world. It is an advantage, not a limitation, which is one reason why we are so eager today to continue uploading our memories to the cloud. We will certainly continue in this direction, as nonbiological systems attain more and more of the capabilities of our biological brains.

My resolution of the dilemma is this: It is not true that You 2 is not you—it *is* you. It is just that there are now two of you. That's not so bad—if you think you are a good thing, then two of you is even better.

What I believe will actually happen is that we will continue on the path of the gradual replacement and augmentation scenario until ultimately most of our thinking will be in the cloud. My leap of faith on identity is that identity is preserved through continuity of the pattern of information that makes us us. Continuity does allow for continual change, so whereas I am somewhat different than I was yesterday, I nonetheless have the same identity. However, the continuity of the pattern that constitutes my identity is not substrate-dependent. Biological substrates are wonderful—they have gotten us very far—but we are creating a more capable and durable substrate for very good reasons.

CHAPTER 10

THE LAW OF ACCELERATING RETURNS
APPLIED TO THE BRAIN

And though man should remain, in some respects, the higher creature, is not this in accordance with the practice of nature, which allows
superiority in some things to animals which have, on the whole, been
long surpassed? Has she not allowed the ant and the bee to retain
superiority over man in the organization of their communities and
social arrangements, the bird in traversing the air, the fish in swim-
ming, the horse in strength and fleetness, and the dog in self-sacrifice?

—Samuel Butler, 1871

There was a time, when the earth was to all appearance utterly desti-
tute both of animal and vegetable life, and when according to the
opinion of our best philosophers it was simply a hot round ball with
a crust gradually cooling. Now if a human being had existed while
the earth was in this state and had been allowed to see it as though it
were some other world with which he had no concern, and if at the
same time he were entirely ignorant of all physical science, would he
not have pronounced it impossible that creatures possessed of anything like consciousness should be evolved from the seeming cinder
which he was beholding? Would he not have denied that it contained

any potentiality of consciousness? Yet in the course of time consciousness came. Is it not possible then that there may be even yet new channels dug out for consciousness, though we can detect no signs of them at present?

—Samuel Butler, 1871

When we reflect upon the manifold phases of life and consciousness which have been evolved already, it would be rash to say that no others can be developed, and that animal life is the end of all things. There was a time when fire was the end of all things: another when rocks and water were so.

—Samuel Butler, 1871

There is no security against the ultimate development of mechanical consciousness, in the fact of machines possessing little consciousness now. A mollusk has not much consciousness. Reflect upon the extraordinary advance which machines have made during the last few hundred years, and note how slowly the animal and vegetable kingdoms are advancing. The more highly organized machines are creatures not so much of yesterday, as of the last five minutes, so to speak, in comparison with past time. Assume for the sake of argument that conscious beings have existed for some twenty million years: see what strides machines have made in the last thousand! May not the world last twenty million years longer? If so, what will they not in the end become?

—Samuel Butler, 1871

My core thesis, which I call the law of accelerating returns (LOAR), is that fundamental measures of information technology follow predictable and exponential trajectories, belying the conventional wisdom that "you can't predict the future." There are still many things—which project, company, or technical standard will prevail in the marketplace, when peace will come to the Middle East—that remain unknowable, but the underlying price/performance and capacity of information has nonetheless proven to be remarkably predictable. Surprisingly, these trends are unperturbed by conditions such as war or peace and prosperity or recession.

A primary reason that evolution created brains was to predict the future. As one of our ancestors walked through the savannas thousands of years ago, she might have noticed that an animal was progressing toward a route that she was taking. She would predict that if she stayed on course, their paths would intersect. Based on this, she decided to head in another direction, and her foresight proved valuable to survival.

But such built-in predictors of the future are linear, not exponential, a quality that stems from the linear organization of the neocortex. Recall that the neocortex is constantly making predictions—what letter and word we will see next, whom we expect to see as we round the corner, and so on. The neocortex is organized with linear sequences of steps in each pattern, which means that exponential thinking does not come naturally to us. The cerebellum also uses linear predictions. When it helps us to catch a fly ball it is making a linear prediction about where the ball will be in our visual field of view and where our gloved hand should be in our visual field of view to catch it.

As I have pointed out, there is a dramatic difference between linear and exponential progressions (forty steps linearly is forty, but exponentially is a trillion), which accounts for why my predictions stemming from the law of accelerating returns seem surprising to many observers at first. We have to train ourselves to think exponentially. When it comes to information technologies, it is the right way to think.

The quintessential example of the law of accelerating returns is the perfectly smooth, doubly exponential growth of the price/performance of computation, which has held steady for 110 years through two world wars, the Great Depression, the Cold War, the collapse of the Soviet Union, the reemergence of China, the recent financial crisis, and all of the other notable events of the late nineteenth, twentieth, and early twenty-first centuries. Some people refer to this phenomenon as "Moore's law," but that is a misconception. Moore's law—which states that you can place twice as many components on an integrated circuit every two years, and they run faster because they are smaller—is just one paradigm among many. It was in fact the fifth, not the first, paradigm to bring exponential growth to the price/performance of computing.

The exponential rise of computation started with the 1890 U.S. census (the first to be automated) using the first paradigm of electromechanical calculation, decades before Gordon Moore was even born. In *The Singularity Is Near* I provide this graph through 2002, and here I update it through 2009 (see the graph on page 257 titled "Exponential Growth of Computing for 110 Years"). The smoothly predictable trajectory has continued, even through the recent economic downturn.

Computation is the most important example of the law of accelerating returns, because of the amount of data we have for it, the ubiquity of computation, and its key role in ultimately revolutionizing everything we care about. But it is far from the only example. Once a technology becomes an information technology, it becomes subject to the LOAR.

Biomedicine is becoming the most significant recent area of technology and industry to be transformed in this way. Progress in medicine has historically been based on accidental discoveries, so progress during the earlier era was linear, not exponential. This has nevertheless been beneficial: Life expectancy has grown from twenty-three years as of a thousand years ago, to thirty-seven years as of two hundred years ago, to close to eighty years today. With the gathering of the software of life—the genome—medicine and human biology have become an information technology. The human genome project itself was perfectly exponential, with the amount of genetic data doubling and the cost per base pair coming down by half each year

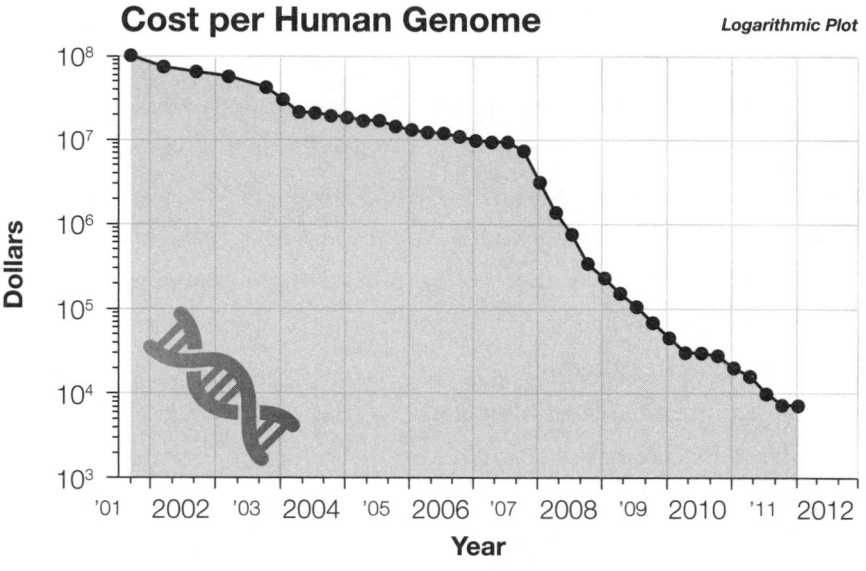

The cost of sequencing a human-sized genome.[1]

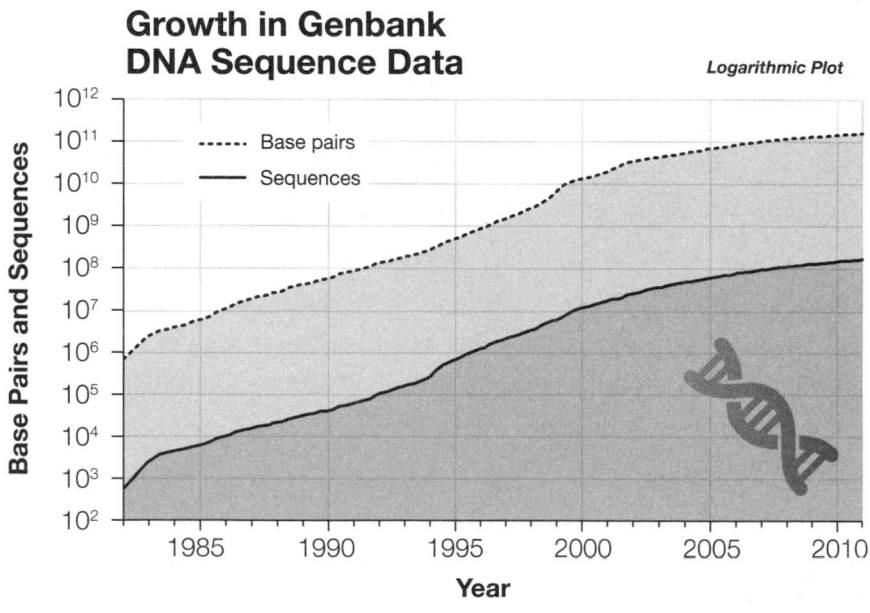

The amount of genetic data sequenced in the world each year.[2]

since the project was initiated in 1990.[3] (All the graphs in this chapter have been updated since *The Singularity Is Near* was published.)

We now have the ability to design biomedical interventions on computers and to test them on biological simulators, the scale and precision of which are also doubling every year. We can also update our own obsolete software: RNA interference can turn genes off, and new forms of gene therapy can add new genes, not just to a newborn but to a mature individual. The advance of genetic technologies also affects the brain reverse-engineering project, in that one important aspect of it is understanding how genes control brain functions such as creating new connections to reflect recently added cortical knowledge. There are many other manifestations of this integration of biology and information technology, as we move beyond genome sequencing to genome synthesizing.

Another information technology that has seen smooth exponential growth is our ability to communicate with one another and transmit vast repositories of human knowledge. There are many ways to measure this phenomenon. Cooper's law, which states that the total bit capacity of wireless communications in a given amount of radio spectrum doubles every thirty months, has held true from the time Guglielmo Marconi used the wireless telegraph for Morse code transmissions in 1897 to today's 4G communications technologies.[4] According to Cooper's law, the amount of information that can be transmitted over a given amount of radio spectrum has been doubling every two and a half years for more than a century. Another example is the number of bits per second transmitted on the Internet, which is doubling every one and a quarter years.[5]

The reason I became interested in trying to predict certain aspects of technology is that I realized about thirty years ago that the key to becoming successful as an inventor (a profession I adopted when I was five years old) was timing. Most inventions and inventors fail not because the gadgets themselves don't work, but because their timing is wrong, appearing either before all of the enabling factors are in place or too late, having missed the window of opportunity.

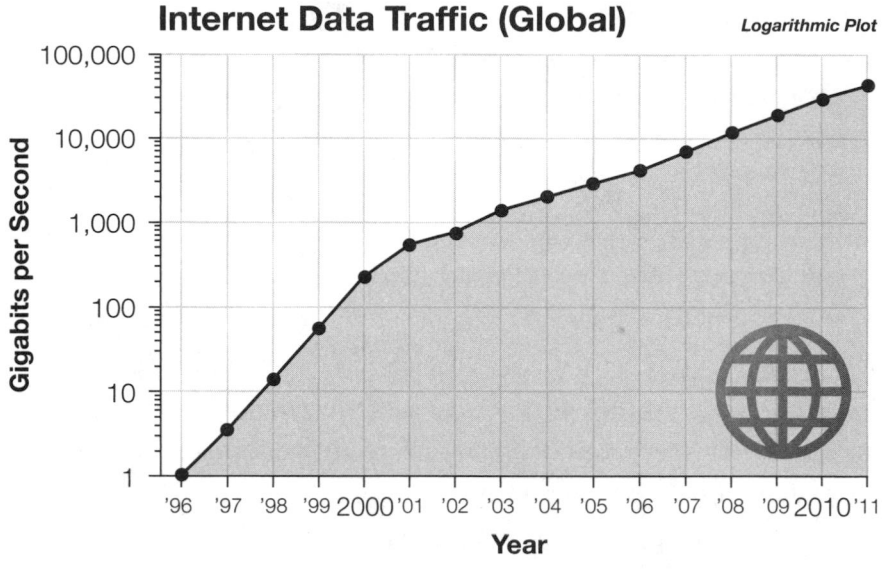

The international (country-to-country) bandwidth dedicated to the Internet for the world.[6]

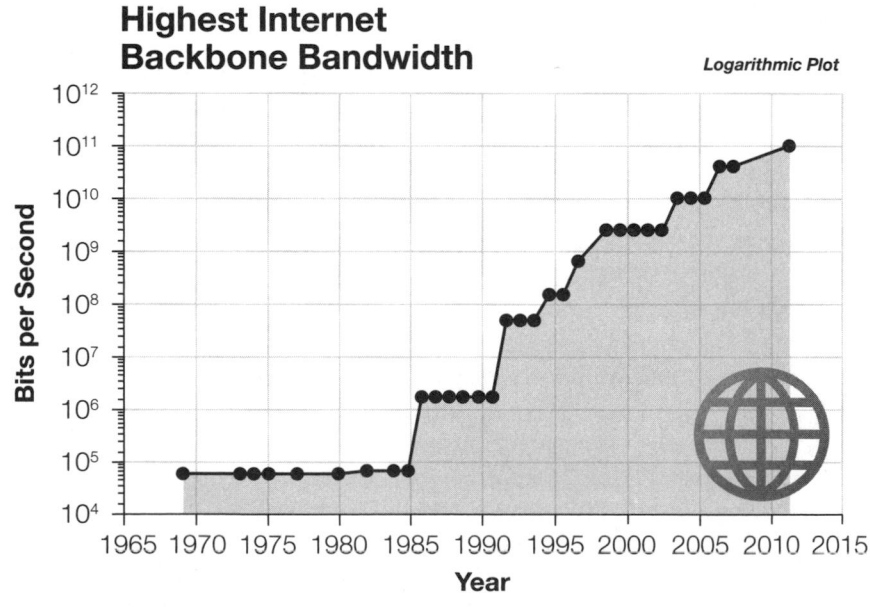

The highest bandwidth (speed) of the Internet backbone.[7]

Being an engineer, about three decades ago I started to gather data on measures of technology in different areas. When I began this effort, I did not expect that it would present a clear picture, but I did hope that it would provide some guidance and enable me to make educated guesses. My goal was—and still is—to time my own technology efforts so that they will be appropriate for the world that exists when I complete a project—which I realized would be very different from the world that existed when I started.

Consider how much and how quickly the world has changed only recently. Just a few years ago, people did not use social networks (Facebook, for example, was founded in 2004 and had 901 million monthly active users at the end of March 2012),[8] wikis, blogs, or tweets. In the 1990s most people did not use search engines or cell phones. Imagine the world without them. That seems like ancient history but was not so long ago. The world will change even more dramatically in the near future.

In the course of my investigation, I made a startling discovery: If a technology is an information technology, the basic measures of price/performance and capacity (per unit of time or cost, or other resource) follow amazingly precise exponential trajectories.

These trajectories outrun the specific paradigms they are based on (such as Moore's law). But when one paradigm runs out of steam (for example, when engineers were no longer able to reduce the size and cost of vacuum tubes in the 1950s), it creates research pressure to create the next paradigm, and so another S-curve of progress begins.

The exponential portion of that next S-curve for the new paradigm then continues the ongoing exponential of the information technology measure. Thus vacuum tube–based computing in the 1950s gave way to transistors in the 1960s, and then to integrated circuits and Moore's law in the late 1960s, and beyond. Moore's law, in turn, will give way to three-dimensional computing, the early examples of which are already in place. The reason why information technologies are able to consistently transcend the limitations of any particular paradigm is that the resources required to compute or remember or transmit a bit of information are vanishingly small.

We might wonder, are there fundamental limits to our ability to compute and transmit information, regardless of paradigm? The answer is yes, based on our current understanding of the physics of computation. Those limits, however, are not very limiting. Ultimately we can expand our intelligence trillions-fold based on molecular computing. By my calculations, we will reach these limits late in this century.

It is important to point out that not every exponential phenomenon is an example of the law of accelerating returns. Some observers misconstrue the LOAR by citing exponential trends that are not information-based: For example, they point out, men's shavers have gone from one blade to two to four, and then ask, where are the eight-blade shavers? Shavers are not (yet) an information technology.

In *The Singularity Is Near,* I provide a theoretical examination, including (in the appendix to that book) a mathematical treatment of why the LOAR is so remarkably predictable. Essentially, we always use the latest technology to create the next. Technologies build on themselves in an exponential manner, and this phenomenon is readily measurable if it involves an information technology. In 1990 we used the computers and other tools of that era to create the computers of 1991; in 2012 we are using current information tools to create the machines of 2013 and 2014. More broadly speaking, this acceleration and exponential growth applies to any process in which patterns of information evolve. So we see acceleration in the pace of biological evolution, and similar (but much faster) acceleration in technological evolution, which is itself an outgrowth of biological evolution.

I now have a public track record of more than a quarter of a century of predictions based on the law of accelerating returns, starting with those presented in *The Age of Intelligent Machines,* which I wrote in the mid-1980s. Examples of accurate predictions from that book include: the emergence in the mid- to late 1990s of a vast worldwide web of communications tying together people around the world to one another and to all human knowledge; a great wave of democratization emerging from this decentralized communication network, sweeping away the Soviet Union; the defeat of the world chess champion by 1998; and many others.

I described the law of accelerating returns, as it is applied to computation, extensively in *The Age of Spiritual Machines,* where I provided a century of data showing the doubly exponential progression of the price/performance of computation through 1998. It is updated through 2009 below.

I recently wrote a 146-page review of the predictions I made in *The Age of Intelligent Machines, The Age of Spiritual Machines,* and *The Singularity Is Near.* (You can read the essay here by going to the link in this endnote.)[9] *The Age of Spiritual Machines* included hundreds of predictions for specific decades (2009, 2019, 2029, and 2099). For example, I made 147 predictions for 2009 in *The Age of Spiritual Machines,* which I wrote in the 1990s. Of these, 115 (78 percent) are entirely correct as of the end of 2009; the predictions that were concerned with basic measurements of the capacity and price/performance of information technologies were particularly accurate.

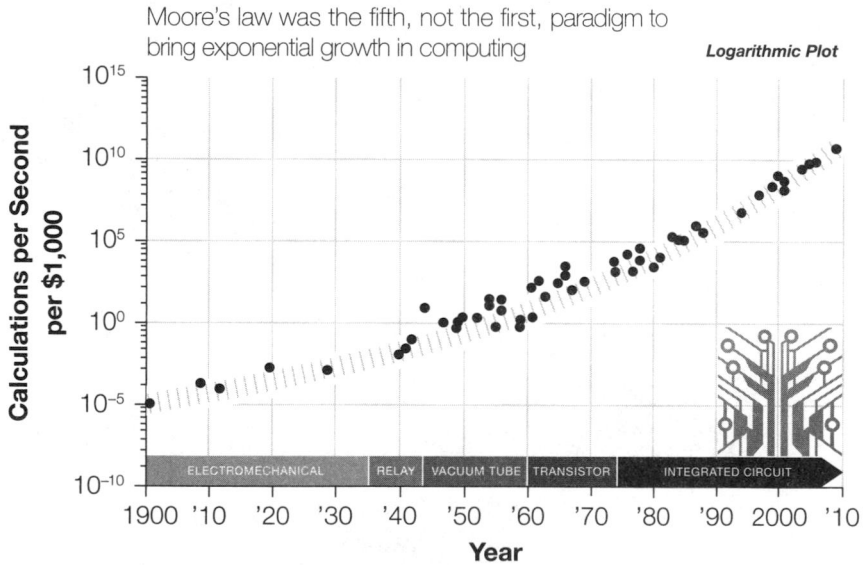

**Exponential Growth
of Computing for 110 Years**

Moore's law was the fifth, not the first, paradigm to bring exponential growth in computing

Logarithmic Plot

Calculations per second per (constant) thousand dollars of different computing devices.[10]

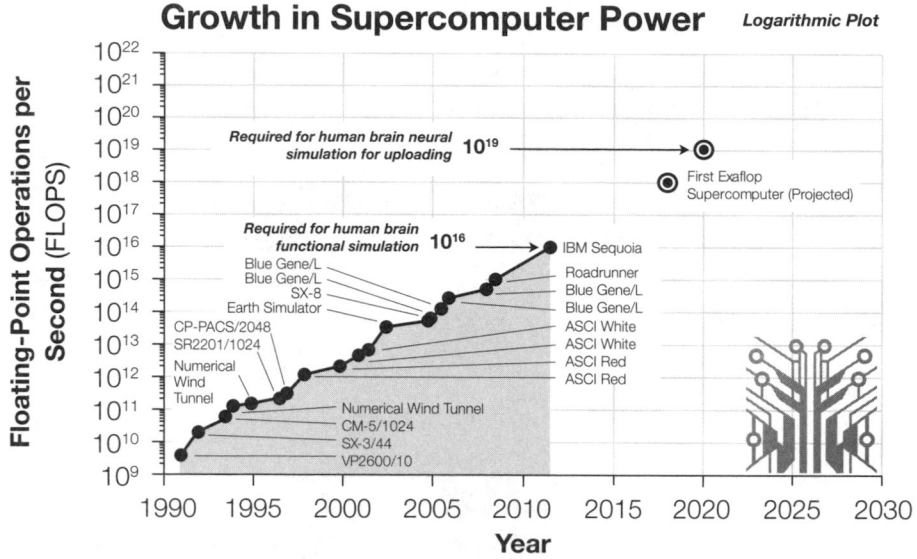

Floating-point operations per second of different supercomputers.[11]

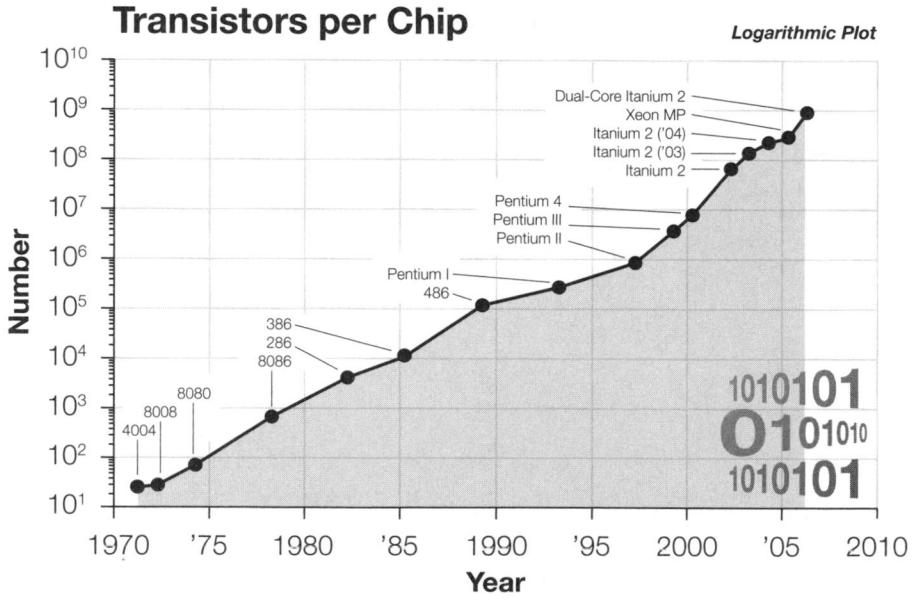

Transistors per chip for different Intel processors.[12]

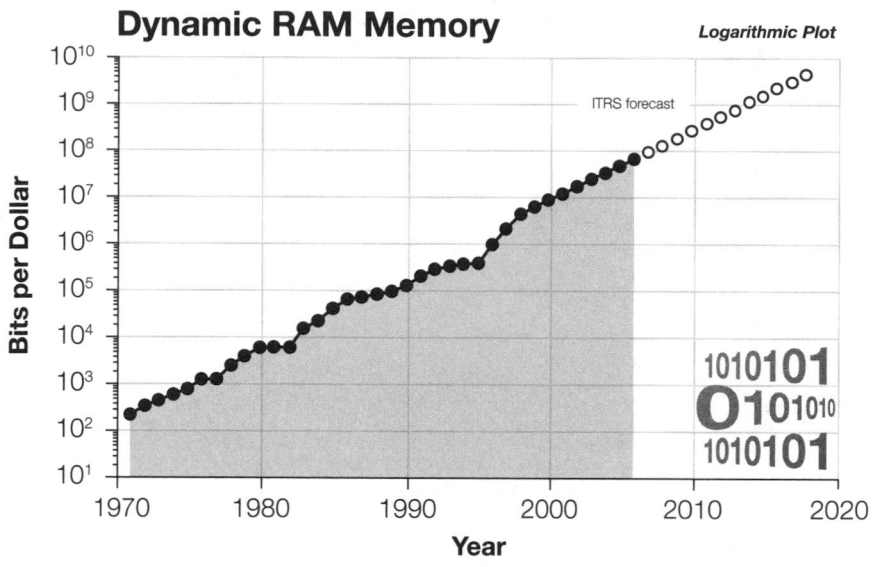

Bits per dollar for dynamic random access memory chips.[13]

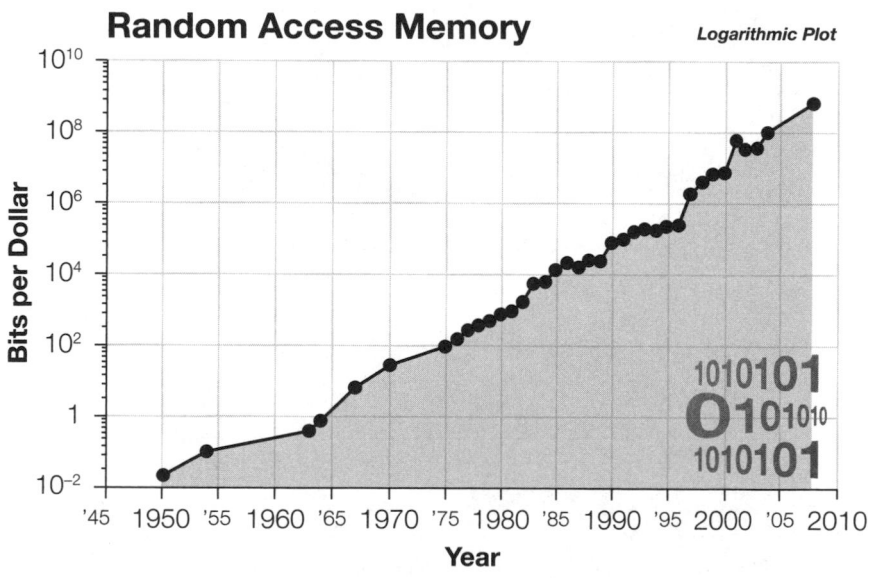

Bits per dollar for random access memory chips.[14]

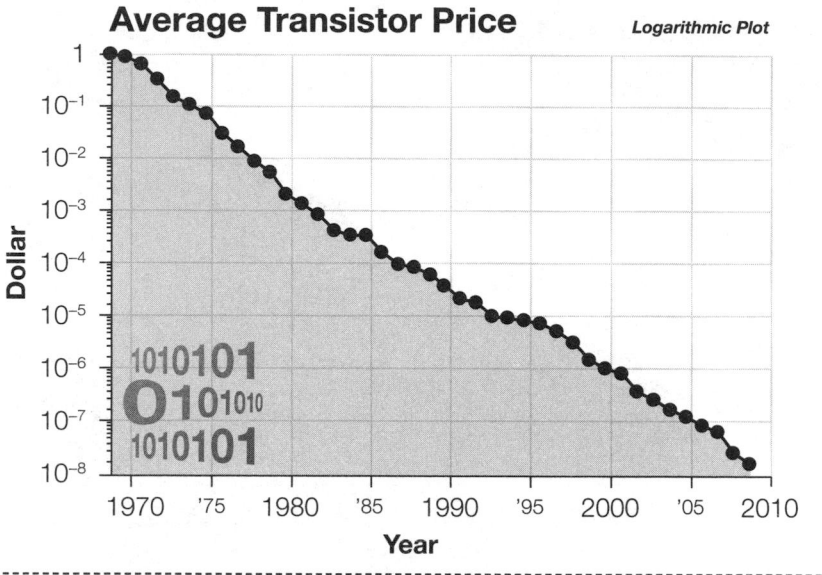

The average price per transistor in dollars.[15]

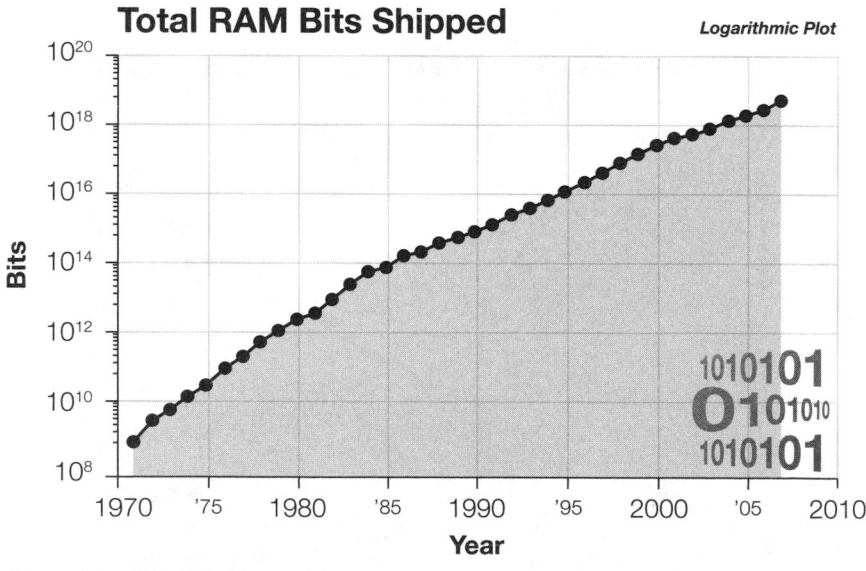

The total number of bits of random access memory shipped each year.[16]

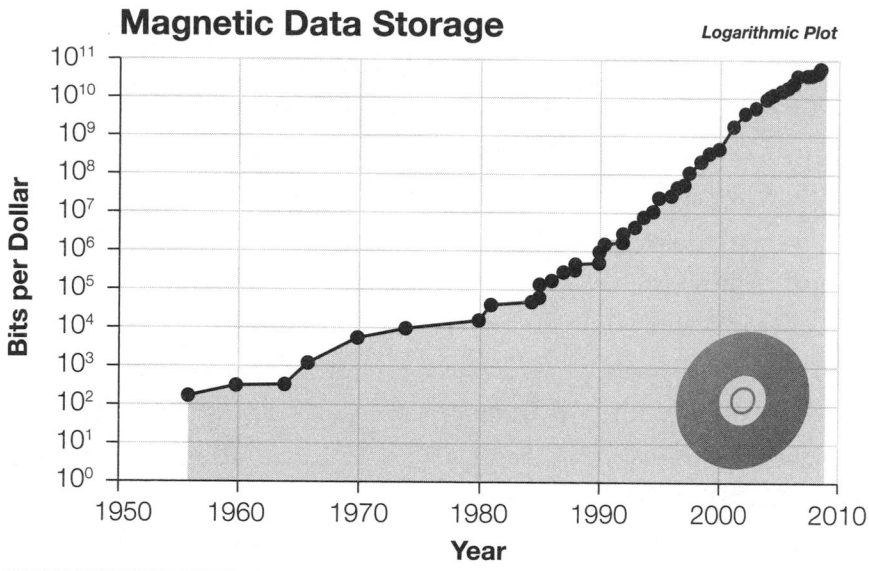

Magnetic Data Storage

Logarithmic Plot

Bits per dollar (in constant 2000 dollars) for magnetic data storage.[17]

Another 12 (8 percent) are "essentially correct." A total of 127 predictions (86 percent) are correct or essentially correct. (Since the predictions were made specific to a given decade, a prediction for 2009 was considered "essentially correct" if it came true in 2010 or 2011.) Another 17 (12 percent) are partially correct, and 3 (2 percent) are wrong.

Even the predictions that were "wrong" were not all wrong. For example, I judged my prediction that we would have self-driving cars to be wrong, even though Google has demonstrated self-driving cars, and even though in October 2010 four driverless electric vans successfully concluded a 13,000-kilometer test drive from Italy to China.[18] Experts in the field currently predict that these technologies will be routinely available to consumers by the end of this decade.

Exponentially expanding computational and communication technologies all contribute to the project to understand and re-create the methods of the human brain. This effort is not a single organized project but rather the result of a great many diverse projects, including detailed modeling of constituents of the brain ranging from individual neurons to the entire neocortex, the mapping of the "connectome" (the neural connections in the brain), simulations of brain regions, and many others. All of these have been scaling up exponentially. Much of the evidence presented in this book has only become available recently—for example, the 2012 Wedeen study discussed in chapter 4 that showed the very orderly and "simple" (to quote the researchers) gridlike pattern of the connections in the neocortex. The researchers in that study acknowledge that their insight (and images) only became feasible as the result of new high-resolution imaging technology.

Brain scanning technologies are improving in resolution, spatial and temporal, at an exponential rate. Different types of brain scanning methods being pursued range from completely noninvasive methods that can be used with humans to more invasive or destructive methods on animals.

MRI (magnetic resonance imaging), a noninvasive imaging technique

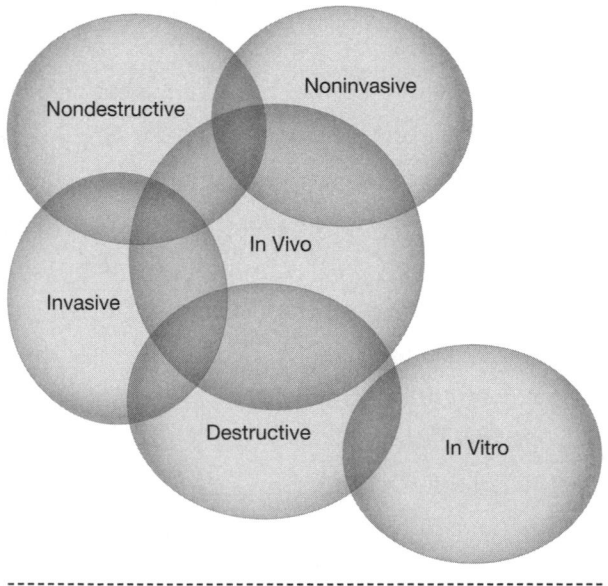

A Venn diagram of brain imaging methods.[19]

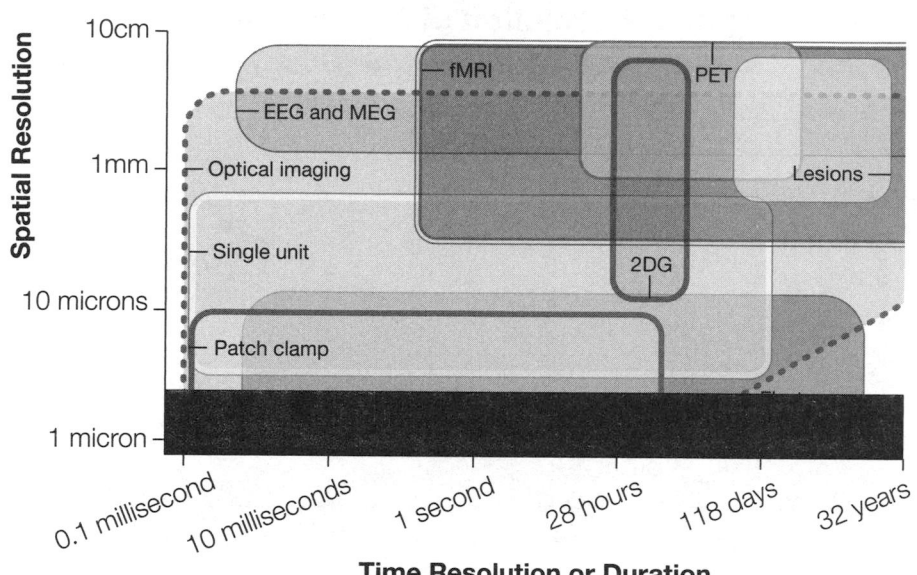

Tools for imaging the brain.[20]

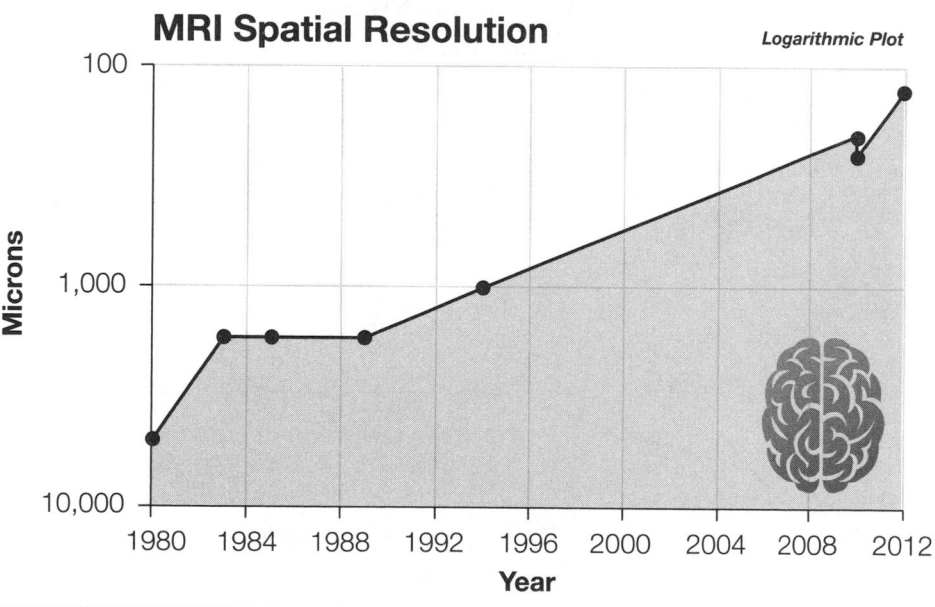

MRI spatial resolution in microns.[21]

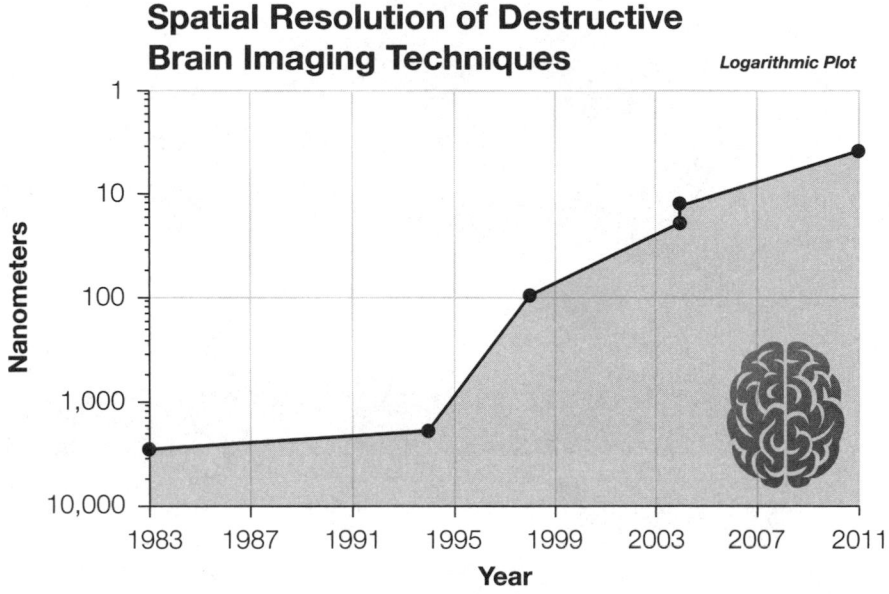

Spatial resolution of destructive imaging techniques.[22]

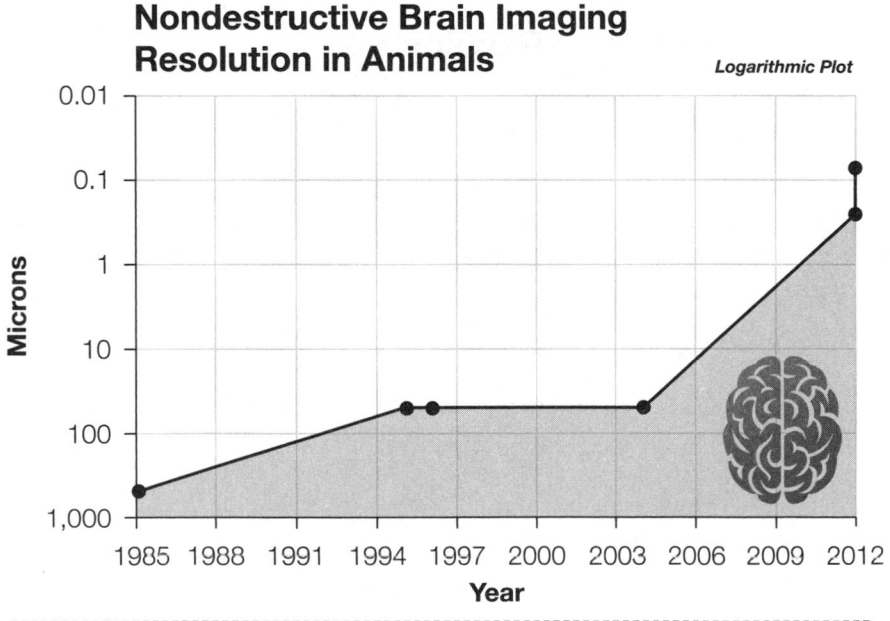

Spatial resolution of nondestructive imaging techniques in animals.[23]

with relatively high temporal resolution, has steadily improved at an exponential pace, to the point that spatial resolutions are now close to 100 microns (millionths of a meter).

Destructive imaging, which is performed to collect the connectome (map of all interneuronal connections) in animal brains, has also improved at an exponential pace. Current maximum resolution is around four nanometers, which is sufficient to see individual connections.

Artificial intelligence technologies such as natural-language-understanding systems are not necessarily designed to emulate theorized principles of brain function, but rather for maximum effectiveness. Given this, it is notable that the techniques that have won out are consistent with the principles I have outlined in this book: self-organizing, hierarchical recognizers of invariant self-associative patterns with redundancy and up-and-down predictions. These systems are also scaling up exponentially, as Watson has demonstrated.

A primary purpose of understanding the brain is to expand our toolkit of techniques to create intelligent systems. Although many AI researchers may not fully appreciate this, they have already been deeply influenced by our knowledge of the principles of the operation of the brain. Understanding the brain also helps us to reverse brain dysfunctions of various kinds. There is, of course, another key goal of the project to reverse-engineer the brain: understanding who we are.

CHAPTER 11

OBJECTIONS

If a machine can prove indistinguishable from a human, we should award it the respect we would to a human—we should accept that it has a mind.

—Stevan Harnad

The most significant source of objection to my thesis on the law of accelerating returns and its application to the amplification of human intelligence stems from the linear nature of human intuition. As I described earlier, each of the several hundred million pattern recognizers in the neocortex processes information sequentially. One of the implications of this organization is that we have linear expectations about the future, so critics apply their linear intuition to information phenomena that are fundamentally exponential.

I call objections along these lines "criticism from incredulity," in that exponential projections seem incredible given our linear predilection, and they take a variety of forms. Microsoft cofounder Paul Allen (born in 1953) and his colleague Mark Greaves recently articulated several of them in an essay titled "The Singularity Isn't Near" published in *Technology Review* magazine.[1] While my response here is to Allen's particular critiques, they represent a typical range of objections to the arguments I've made, especially with regard to the brain. Although Allen references *The Singularity Is Near* in the title of his essay, his only citation in the piece is to an essay I wrote in 2001 ("The Law of Accelerating Returns"). Moreover, his article does not acknowledge or respond to arguments I actually make in the book. Unfortunately, I find this often to be the case with critics of my work.

When *The Age of Spiritual Machines* was published in 1999, augmented later by the 2001 essay, it generated several lines of criticism, such as: *Moore's law will come to an end; hardware capability may be expanding exponentially but software is stuck in the mud; the brain is too complicated; there are capabilities in the brain that inherently cannot be replicated in software;* and several others. One of the reasons I wrote *The Singularity Is Near* was to respond to those critiques.

I cannot say that Allen and similar critics would necessarily have been convinced by the arguments I made in that book, but at least he and others could have responded to what I actually wrote. Allen argues that "the Law of Accelerating Returns (LOAR) . . . is not a physical law." I would point out that most scientific laws are not physical laws, but result from the emergent properties of a large number of events at a lower level. A classic example is the laws of thermodynamics (LOT). If you look at the mathematics underlying the LOT, it models each particle as following a random walk, so by definition we cannot predict where any particular particle will be at any future time. Yet the overall properties of the gas are quite predictable to a high degree of precision, according to the *laws* of thermodynamics. So it is with the law of accelerating returns: Each technology project and contributor is unpredictable, yet the overall trajectory, as quantified by basic measures of price/performance and capacity, nonetheless follows a remarkably predictable path.

If computer technology were being pursued by only a handful of researchers, it would indeed be unpredictable. But it's the product of a sufficiently dynamic system of competitive projects that a basic measure of its price/performance, such as calculations per second per constant dollar, follows a very smooth exponential path, dating back to the 1890 American census as I noted in the previous chapter. While the theoretical basis for the LOAR is presented extensively in *The Singularity Is Near,* the strongest case for it is made by the extensive empirical evidence that I and others present.

Allen writes that "these 'laws' work until they don't." Here he is confusing paradigms with the ongoing trajectory of a basic area of information technology. If we were examining, for example, the trend of creating ever smaller vacuum tubes—the paradigm for improving computation in the 1950s—it's true that it continued until it didn't. But as the end of this particular paradigm became clear, research pressure grew for the next paradigm. The technology of transistors kept the underlying trend of the exponential growth of price/performance of computation going, and that led to the fifth paradigm (Moore's law) and the continual compression of features on integrated circuits. There have been regular predictions that Moore's law will come to an end. The semiconductor industry's "International Technology Roadmap for Semiconductors" projects seven-nanometer features by the early 2020s.[2] At that point key features will be the width of thirty-five carbon atoms, and it will be difficult to continue shrinking them any farther. However, Intel and other chip makers are already taking the first steps toward the sixth paradigm, computing in three dimensions, to continue exponential improvement in price/performance. Intel projects that three-dimensional chips will be mainstream by the teen years; three-dimensional transistors and 3-D memory chips have already been introduced. This sixth paradigm will keep the LOAR going with regard to computer price/performance to a time later in this century when a thousand dollars' worth of computation will be trillions of times more powerful than the human brain.[3] (It appears that Allen and I are at least in agreement on what level of computation is required to functionally simulate the human brain.)[4]

Allen then goes on to give the standard argument that software is not progressing in the same exponential manner as hardware. In *The Singularity Is Near* I addressed this issue at length, citing different methods of measuring complexity and capability in software that do demonstrate a similar exponential growth.[5] One recent study ("Report to the President and Congress, Designing a Digital Future: Federally Funded Research and Development in Networking and Information Technology," by the President's Council of Advisors on Science and Technology) states the following:

> Even more remarkable—and even less widely understood—is that in many areas, *performance gains due to improvements in algorithms have vastly exceeded even the dramatic performance gains due to increased processor speed.* The algorithms that we use today for speech recognition, for natural language translation, for chess playing, for logistics planning, have evolved remarkably in the past decade. . . . Here is just one example, provided by Professor Martin Grötschel of Konrad-Zuse-Zentrum für Informationstechnik Berlin. Grötschel, an expert in optimization, observes that a benchmark production planning model solved using linear programming would have taken 82 years to solve in 1988, using the computers and the linear programming algorithms of the day. Fifteen years later—in 2003—this same model could be solved in roughly 1 minute, an improvement by a factor of roughly 43 million. Of this, a factor of roughly 1,000 was due to increased processor speed, whereas a factor of roughly 43,000 was due to improvements in algorithms! Grötschel also cites an algorithmic improvement of roughly 30,000 for mixed integer programming between 1991 and 2008. The design and analysis of algorithms, and the study of the inherent computational complexity of problems, are fundamental subfields of computer science.

Note that the linear programming that Grötschel cites above as having benefited from an improvement in performance of 43 million to 1 is the mathematical technique that is used to optimally assign resources in a hierarchical memory system such as HHMM that I discussed earlier. I cite many other similar examples like this in *The Singularity Is Near.*[6]

Regarding AI, Allen is quick to dismiss IBM's Watson, an opinion shared by many other critics. Many of these detractors don't know anything about Watson other than the fact that it is software running on a computer (albeit a parallel one with 720 processor cores). Allen writes that systems such as Watson "remain brittle, their performance boundaries are rigidly set by their internal assumptions and defining algorithms, they cannot generalize, and they frequently give nonsensical answers outside of their specific areas."

First of all, we could make a similar observation about humans. I would also point out that Watson's "specific areas" include *all* of Wikipedia plus many other knowledge bases, which hardly constitute a narrow focus. Watson deals with a vast range of human knowledge and is capable of dealing with subtle forms of language, including puns, similes, and metaphors in virtually all fields of human endeavor. It's not perfect, but neither are humans, and it was good enough to be victorious on *Jeopardy!* over the best human players.

Allen argues that Watson was assembled by the scientists themselves, building each link of narrow knowledge in specific areas. This is simply not true. Although a few areas of Watson's data were programmed directly, Watson acquired the significant majority of its knowledge on its own by reading natural-language documents such as Wikipedia. That represents its key strength, as does its ability to understand the convoluted language in *Jeopardy!* queries (answers in search of a question).

As I mentioned earlier, much of the criticism of Watson is that it works through statistical probabilities rather than "true" understanding. Many readers interpret this to mean that Watson is merely gathering statistics on word sequences. The term "statistical information" in the case of Watson actually refers to distributed coefficients and symbolic connections in self-organizing methods such as hierarchical hidden Markov models. One could just as easily dismiss the distributed neurotransmitter concentrations and redundant connection patterns in the human cortex as "statistical information." Indeed we resolve ambiguities in much the same way that Watson does—by considering the likelihood of different interpretations of a phrase.

Allen continues, "Every structure [in the brain] has been precisely shaped by millions of years of evolution to do a particular thing, whatever it might be. It is not like a computer, with billions of identical transistors in regular memory arrays that are controlled by a CPU with a few different elements. In the brain every individual structure and neural circuit has been individually refined by evolution and environmental factors."

This contention that every structure and neural circuit in the brain is unique and there by design is simply impossible, for it would mean that the blueprint of the brain would require hundreds of trillions of bytes of information. The brain's structural plan (like that of the rest of the body) is contained in the genome, and the brain itself cannot contain more design information than the genome. Note that epigenetic information (such as the peptides controlling gene expression) does not appreciably add to the amount of information in the genome. Experience and learning do add significantly to the amount of information contained in the brain, but the same can be said of AI systems like Watson. I show in *The Singularity Is Near* that, after lossless compression (due to massive redundancy in the genome), the amount of design information in the genome is about 50 million bytes, roughly half of which (that is, about 25 million bytes) pertains to the brain.[7] That's not simple, but it is a level of complexity we can deal with and represents less complexity than many software systems in the modern world. Moreover much of the brain's 25 million bytes of genetic design information pertain to the biological requirements of neurons, not to their information-processing algorithms.

How do we arrive at on the order of 100 to 1,000 trillion connections in the brain from only tens of millions of bytes of design information? Obviously, the answer is through massive redundancy. Dharmendra Modha, manager of Cognitive Computing for IBM Research, writes that "neuroanatomists have not found a hopelessly tangled, arbitrarily connected network, completely idiosyncratic to the brain of each individual, but instead a great deal of repeating structure within an individual brain and a great deal of homology across species. . . . The astonishing natural reconfigurability gives hope that the core algorithms of neurocomputation are independent of the specific sensory or motor modalities and that much of the observed variation in cortical structure across areas represents a refinement of a canonical circuit; it is indeed this canonical circuit we wish to reverse engineer."[8]

Allen argues in favor of an inherent "complexity brake that would necessarily limit progress in understanding the human brain and replicating its capabilities," based on his notion that each of the approximately 100 to 1,000 trillion connections in the human brain is there by explicit design. His "complexity brake" confuses the forest with the trees. If you want to understand, model, simulate, and re-create a pancreas, you don't need to re-create or simulate every organelle in every pancreatic islet cell. You would want instead to understand one islet cell, then abstract its basic functionality as it pertains to insulin control, and then extend that to a large group of such cells. This algorithm is well understood with regard to islet cells. There are now artificial pancreases that utilize this functional model being tested. Although there is certainly far more intricacy and variation in the brain than in the massively repeated islet cells of the pancreas, there is nonetheless massive repetition of functions, as I have described repeatedly in this book.

Critiques along the lines of Allen's also articulate what I call the "scientist's pessimism." Researchers working on the next generation of a technology or of modeling a scientific area are invariably struggling with that immediate set of challenges, so if someone describes what the technology will look like in ten generations, their eyes glaze over. One of the pioneers of integrated circuits was recalling for me recently the struggles to go from 10-micron (10,000 nanometers) feature sizes to 5-micron (5,000 nanometers) features over thirty years ago. The scientists were cautiously confident of reaching this goal, but when people predicted that someday we would actually have circuitry with feature sizes under 1 micron

(1,000 nanometers), most of them, focused on their own goal, thought that was too wild to contemplate. Objections were made regarding the fragility of circuitry at that level of precision, thermal effects, and so on. Today Intel is starting to use chips with 22-nanometer gate lengths.

We witnessed the same sort of pessimism with respect to the Human Genome Project. Halfway through the fifteen-year effort, only 1 percent of the genome had been collected, and critics were proposing basic limits on how quickly it could be sequenced without destroying the delicate genetic structures. But thanks to the exponential growth in both capacity and price/performance, the project was finished seven years later. The project to reverse-engineer the human brain is making similar progress. It is only recently, for example, that we have reached a threshold with noninvasive scanning techniques so that we can see individual interneuronal connections forming and firing in real time. Much of the evidence I have presented in this book was dependent on such developments and has only recently been available.

Allen describes my proposal about reverse-engineering the human brain as simply scanning the brain to understand its fine structure and then simulating an entire brain "bottom up" without comprehending its information-processing methods. This is not my proposition. We do need to understand in detail how individual types of neurons work, and then gather information about how functional modules are connected. The functional methods that are derived from this type of analysis can then guide the development of intelligent systems. Basically, we are looking for biologically inspired methods that can accelerate work in AI, much of which has progressed without significant insight as to how the brain performs similar functions. From my own work in speech recognition, I know that our work was greatly accelerated when we gained insights as to how the brain prepares and transforms auditory information.

The way that the massively redundant structures in the brain differentiate is through learning and experience. The current state of the art in AI does in fact enable systems to also learn from their own experience. The

Google self-driving cars learn from their own driving experience as well as from data from Google cars driven by human drivers; Watson learned most of its knowledge by reading on its own. It is interesting to note that the methods deployed today in AI have evolved to be mathematically very similar to the mechanisms in the neocortex.

Another objection to the feasibility of "strong AI" (artificial intelligence at human levels and beyond) that is often raised is that the human brain makes extensive use of analog computing, whereas digital methods inherently cannot replicate the gradations of value that analog representations can embody. It is true that one bit is either on or off, but multiple-bit words easily represent multiple gradations and can do so to any desired degree of accuracy. This is, of course, done all the time in digital computers. As it is, the accuracy of analog information in the brain (synaptic strength, for example) is only about one level within 256 levels that can be represented by eight bits.

In chapter 9 I cited Roger Penrose and Stuart Hameroff's objection, which concerned microtubules and quantum computing. Recall that they claim that the microtubule structures in neurons are doing quantum computing, and since it is not possible to achieve that in computers, the human brain is fundamentally different and presumably better. As I argued earlier, there is no evidence that neuronal microtubules are carrying out quantum computation. Humans in fact do a very poor job of solving the kinds of problems that a quantum computer would excel at (such as factoring large numbers). And if any of this proved to be true, there would be nothing barring quantum computing from also being used in our computers.

John Searle is famous for introducing a thought experiment he calls "the Chinese room," an argument I discuss in detail in *The Singularity Is Near*.[9] In short, it involves a man who takes in written questions in Chinese and then answers them. In order to do this, he uses an elaborate rulebook. Searle claims that the man has no true understanding of Chinese and is not "conscious" of the language (as he does not understand the questions or the answers) despite his apparent ability to answer questions in Chinese. Searle compares this to a computer and concludes that a computer that could answer questions in Chinese (essentially passing a Chinese Turing test) would, like the man in the Chinese room, have no real understanding of the language and no consciousness of what it was doing.

There are a few philosophical sleights of hand in Searle's argument. For one thing, the man in this thought experiment is comparable only to the central processing unit (CPU) of a computer. One could say that a CPU has no true understanding of what it is doing, but the CPU is only part of the structure. In Searle's Chinese room, it is the man *with* his rulebook that constitutes the whole system. That system does have an understanding of Chinese; otherwise it would not be capable of convincingly answering questions in Chinese, which would violate Searle's assumption for this thought experiment.

The attractiveness of Searle's argument stems from the fact that it is difficult today to infer true understanding and consciousness in a computer program. The problem with his argument, however, is that you can apply his own line of reasoning to the human brain itself. Each neocortical pattern recognizer—indeed, each neuron and each neuronal component—is following an algorithm. (After all, these are molecular mechanisms that follow natural law.) If we conclude that following an algorithm is inconsistent with true understanding and consciousness, then we would have to also conclude that the human brain does not exhibit these qualities either. You can take John Searle's Chinese room argument and simply substitute "manipulating interneuronal connections and synaptic strengths" for his words "manipulating symbols" and you will have a convincing argument to the effect that human brains cannot truly understand anything.

Another line of argument comes from the nature of nature, which has become a new sacred ground for many observers. For example, New Zealand biologist Michael Denton (born in 1943) sees a profound difference between the design principles of machines and those of biology. Denton writes that natural entities are "self-organizing, . . . self-referential, . . .

self-replicating, . . . reciprocal, . . . self-formative, and . . . holistic."[10] He claims that such biological forms can only be created through biological processes and that these forms are thereby "immutable, . . . impenetrable, and . . . fundamental" realities of existence, and are therefore basically a different philosophical category from machines.

The reality, as we have seen, is that machines can be designed using these same principles. Learning the specific design paradigms of nature's most intelligent entity—the human brain—is precisely the purpose of the brain reverse-engineering project. It is also not true that biological systems are completely "holistic," as Denton puts it, nor, conversely, do machines need to be completely modular. We have clearly identified hierarchies of units of functionality in natural systems, especially the brain, and AI systems are using comparable methods.

It appears to me that many critics will not be satisfied until computers routinely pass the Turing test, but even that threshold will not be clear-cut. Undoubtedly, there will be controversy as to whether claimed Turing tests that have been administered are valid. Indeed, I will probably be among those critics disparaging early claims along these lines. By the time the arguments about the validity of a computer passing the Turing test do settle down, computers will have long since surpassed unenhanced human intelligence.

My emphasis here is on the word "unenhanced," because enhancement is precisely the reason that we are creating these "mind children," as Hans Moravec calls them.[11] Combining human-level pattern recognition with the inherent speed and accuracy of computers will result in very powerful abilities. But this is not an alien invasion of intelligent machines from Mars—we are creating these tools to make ourselves smarter. I believe that most observers will agree with me that this is what is unique about the human species: We build these tools to extend our own reach.

EPILOGUE

The picture's pretty bleak, gentlemen ... The world's climates are changing, the mammals are taking over, and we all have a brain about the size of a walnut.

—Dinosaurs talking, in *The Far Side* by Gary Larson

Intelligence may be defined as the ability to solve problems with limited resources, in which a key such resource is time. Thus the ability to more quickly solve a problem like finding food or avoiding a predator reflects greater power of intellect. Intelligence evolved because it was useful for survival—a fact that may seem obvious, but one with which not everyone agrees. As practiced by our species, it has enabled us not only to dominate the planet but to steadily improve the quality of our lives. This latter point, too, is not apparent to everyone, given that there is a widespread perception today that life is only getting worse. For example, a Gallup poll released on May 4, 2011, revealed that only "44 percent of Americans believed that today's youth will have a better life than their parents."[1]

If we look at the broad trends, not only has human life expectancy quadrupled over the last millennium (and more than doubled in the last two centuries),[2] but per capita GDP (in constant current dollars) has gone from hundreds of dollars in 1800 to thousands of dollars today, with even more pronounced trends in the developed world.[3] Only a handful of democracies existed a century ago, whereas they are the norm today. For a historical perspective on how far we have advanced, I suggest people read Thomas Hobbes's *Leviathan* (1651), in which he describes the "life of man" as "solitary, poor, nasty, brutish, and short." For a modern perspective, the recent book *Abundance* (2012), by X-Prize Foundation founder (and cofounder with me of Singularity University) Peter Diamandis and science writer Steven Kotler, documents the extraordinary ways in which life today has steadily improved in every dimension. Steven Pinker's recent *The Better Angels of Our Nature: Why Violence Has Declined* (2011) painstakingly documents the steady rise of peaceful relations between people and peoples. American lawyer, entrepreneur, and author Martine Rothblatt (born in 1954) documents the steady improvement in civil rights, noting, for example, how in a couple of decades same-sex marriage went from being legally recognized nowhere in the world to being legally accepted in a rapidly growing number of jurisdictions.[4]

A primary reason that people believe that life is getting worse is because our information about the problems of the world has steadily improved. If there is a battle today somewhere on the planet, we experience it almost as if we were there. During World War II, tens of thousands of people might perish in a battle, and if the public could see it at all it was in a grainy newsreel in a movie theater weeks later. During World War I a small elite could read about the progress of the conflict in the newspaper (without pictures). During the nineteenth century there was almost no access to news in a timely fashion for anyone.

The advancement we have made as a species due to our intelligence is reflected in the evolution of our knowledge, which includes our technology and our culture. Our various technologies are increasingly becoming information technologies, which inherently continue to progress in an exponential manner. It is through such technologies that we are able to address the grand challenges of humanity, such as maintaining a healthy environment, providing the resources for a growing population (including energy, food, and water), overcoming disease, vastly extending human longevity, and eliminating poverty. It is only by extending ourselves with intelligent technology that we can deal with the scale of complexity needed to address these challenges.

These technologies are not the vanguard of an intelligent invasion that will compete with and ultimately displace us. Ever since we picked up a stick to reach a higher branch, we have used our tools to extend our reach, both physically and mentally. That we can take a device out of our pocket today and access much of human knowledge with a few keystrokes extends us beyond anything imaginable by most observers only a few decades ago. The "cell phone" (the term is placed in quotes because it is vastly more than a phone) in my pocket is a million times less expensive yet thousands of times more powerful than the computer all the students and professors at MIT shared when I was an undergraduate there. That's a several billion-fold increase in price/performance over the last forty years, an escalation we will see again in the next twenty-five years, when what used to fit in a building, and now fits in your pocket, will fit inside a blood cell.

In this way we will merge with the intelligent technology we are creating. Intelligent nanobots in our bloodstream will keep our biological bodies healthy at the cellular and molecular levels. They will go into our brains noninvasively through the capillaries and interact with our biological neurons, directly extending our intelligence. This is not as futuristic as it may sound. There are already blood cell–sized devices that can cure type I diabetes in animals or detect and destroy cancer cells in the bloodstream. Based on the law of accelerating returns, these technologies will be a billion times more powerful within three decades than they are today.

I already consider the devices I use and the cloud of computing resources to which they are virtually connected as extensions of myself, and feel less than complete if I am cut off from these brain extenders. That is why the one-day strike by Google, Wikipedia, and thousands of other Web sites against the SOPA (Stop Online Piracy Act) on January 18, 2012, was so remarkable: I felt as if part of my brain were going on strike (although I and others did find ways to access these online resources). It was also an impressive demonstration of the political power of these sites as the bill—which looked as if it was headed for ratification—was instantly killed. But more important, it showed how thoroughly we have already outsourced parts of our thinking to the cloud of computing. It is already part of who we are. Once we routinely have intelligent nonbiological intelligence in our brains, this augmentation—and the cloud it is connected to—will continue to grow in capability exponentially.

The intelligence we will create from the reverse-engineering of the brain will have access to its own source code and will be able to rapidly improve itself in an accelerating iterative design cycle. Although there is considerable plasticity in the biological human brain, as we have seen, it does have a relatively fixed architecture, which cannot be significantly modified, as well as a limited capacity. We are unable to increase its 300 million pattern recognizers to, say, 400 million unless we do so nonbiologically. Once we can achieve that, there will be no reason to stop at a particular level of capability. We can go on to make it a billion pattern recognizers, or a trillion.

From quantitative improvement comes qualitative advance. The most important evolutionary advance in *Homo sapiens* was quantitative: the development of a larger forehead to accommodate more neocortex. Greater neocortical capacity enabled this new species to create and contemplate thoughts at higher conceptual levels, resulting in the establishment of all the varied fields of art and science. As we add more neocortex in a nonbiological form, we can expect ever higher qualitative levels of abstraction.

British mathematician Irvin J. Good, a colleague of Alan Turing's, wrote in 1965 that "the first ultraintelligent machine is the last invention that man need ever make." He defined such a machine as one that could surpass the "intellectual activities of any man however clever" and concluded that "since the design of machines is one of these intellectual activities, an ultraintelligent machine could design even better machines; there would then unquestionably be an 'intelligence explosion.'"

The last invention that biological evolution needed to make—the neocortex—is inevitably leading to the last invention that humanity needs to make—truly intelligent machines—and the design of one is inspiring the other. Biological evolution is continuing but technological evolution is moving a million times faster than the former. According to the law of accelerating returns, by the end of this century we will be able to create computation at the limits of what is possible, based on the laws of physics as applied to computation.[5] We call matter and energy organized in this way "computronium," which is vastly more powerful pound per pound than the human brain. It will not just be raw computation but will be infused with intelligent algorithms constituting all of human-machine knowledge. Over time we will convert much of the mass and energy in our tiny corner of the galaxy that is suitable for this purpose to computronium. Then, to keep the law of accelerating returns going, we will need to spread out to the rest of the galaxy and universe.

If the speed of light indeed remains an inexorable limit, then colonizing the universe will take a long time, given that the nearest star system to Earth is four light-years away. If there are even subtle means to circumvent this limit, our intelligence and technology will be sufficiently powerful to exploit them. This is one reason why the recent suggestion that the muons that traversed the 730 kilometers from the CERN accelerator on the Swiss-French border to the Gran Sasso Laboratory in central Italy appeared to be moving faster than the speed of light was such potentially significant news. This particular observation appears to be a false alarm, but there are other possibilities to get around this limit. We do not even need to exceed the speed of light if we can find shortcuts to other apparently faraway places through spatial dimensions beyond the three with which we are familiar. Whether we are able to surpass or otherwise get around the speed of light as a limit will be the key strategic issue for the human-machine civilization at the beginning of the twenty-second century.

Cosmologists argue about whether the world will end in fire (a big crunch to match the big bang) or ice (the death of the stars as they spread out into an eternal expansion), but this does not take into account the power of intelligence, as if its emergence were just an entertaining sideshow to the grand celestial mechanics that now rule the universe. How long will it take for us to spread our intelligence in its nonbiological form throughout the universe? If we can transcend the speed of light—admittedly a big if—for example, by using wormholes through space (which are consistent with our current understanding of physics), it could be achieved within a few centuries. Otherwise, it will likely take much longer. In either scenario, waking up the universe, and then intelligently deciding its fate by infusing it with our human intelligence in its nonbiological form, is our destiny.

NOTES

Introduction

1. Here is one sentence from *One Hundred Years of Solitude* by Gabriel García Márquez:

> Aureliano Segundo was not aware of the singsong until the following day after breakfast when he felt himself being bothered by a buzzing that was by then more fluid and louder than the sound of the rain, and it was Fernanda, who was walking throughout the house complaining that they had raised her to be a queen only to have her end up as a servant in a madhouse, with a lazy, idolatrous, libertine husband who lay on his back waiting for bread to rain down from heaven while she was straining her kidneys trying to keep afloat a home held together with pins where there was so much to do, so much to bear up under and repair from the time God gave his morning sunlight until it was time to go to bed that when she got there her eyes were full of ground glass, and yet no one ever said to her, "Good morning, Fernanda, did you sleep well?," nor had they asked her, even out of courtesy, why she was so pale or why she awoke with purple rings under her eyes in spite of the fact that she expected it, of course, from a family that had always considered her a nuisance, an old rag, a booby painted on the wall, and who were always going around saying things against her behind her back, calling her churchmouse, calling her Pharisee, calling her crafty, and even Amaranta, may she rest in peace, had said aloud that she was one of those people who could not tell their rectums from their ashes, God have mercy, such words, and she had tolerated everything with resignation because of the Holy Father, but she had not been able to tolerate it any more when that evil José Arcadio Segundo

said that the damnation of the family had come when it opened its doors to a stuck-up highlander, just imagine, a bossy highlander, Lord save us, a highlander daughter of evil spit of the same stripe as the highlanders the government sent to kill workers, you tell me, and he was referring to no one but her, the godchild of the Duke of Alba, a lady of such lineage that she made the liver of presidents' wives quiver, a noble dame of fine blood like her, who had the right to sign eleven peninsular names and who was the only mortal creature in that town full of bastards who did not feel all confused at the sight of sixteen pieces of silverware, so that her adulterous husband could die of laughter afterward and say that so many knives and forks and spoons were not meant for a human being but for a centipede, and the only one who could tell with her eyes closed when the white wine was served and on what side and in which glass and when the red wine and on what side and in which glass and not like that peasant of an Amaranta, may she rest in peace, who thought that white wine was served in the daytime and red wine at night, and the only one on the whole coast who could take pride in the fact that she took care of her bodily needs only in golden chamberpots, so that Colonel Aureliano Buendía, may he rest in peace, could have the effrontery to ask her with his Masonic ill humor where she had received that privilege and whether she did not shit shit but shat sweet basil, just imagine, with those very words, and so that Renata, her own daughter, who through an oversight had seen her stool in the bedroom, had answered that even if the pot was all gold and with a coat of arms, what was inside was pure shit, physical shit, and worse even than any other kind because it was stuck-up highland shit, just imagine, her own daughter, so that she never had any illusions about the rest of the family, but in any case she had the right to expect a little more consideration from her husband because, for better or for worse, he was her consecrated spouse, her helpmate, her legal despoiler, who took upon himself of his own free and sovereign will the grave responsibility of taking her away from her paternal home, where she never wanted for or suffered from anything, where she wove funeral wreaths as a pastime, since her godfather had sent a letter with his signature and the stamp of his ring on the sealing wax simply to say that the hands of his goddaughter were not meant for tasks of this world except to play the clavichord, and, nevertheless, her insane husband had taken her from her home with all manner of admonitions and warnings and had brought her to that frying pan of hell where a person could not breathe because of the heat, and before she had completed her Pentecostal fast he had gone off with his wandering trunks and his wastrel's accordion to loaf in adultery with a wretch of whom it was only enough to see her behind, well, that's been said, to see her wiggle her mare's behind in order to guess that she was a, that she was a, just the opposite of her, who was a lady in a palace or a pigsty, at the table or in bed, a lady of breeding, God-fearing, obeying His laws and submissive to His wishes, and with whom he could not perform, naturally, the acrobatics and trampish antics that he did with the other one, who, of course, was ready for anything, like the French matrons, and even worse, if one considers well, because they at least had the honesty to put a red light at their door, swinishness like that, just imagine, and that was all that was needed by the only and beloved daughter of Doña Renata Argote and Don Fernando del Carpio, and especially the latter, an upright man, a fine Christian, a Knight of the Order of the Holy Sepulcher, those who receive direct from God the privilege of remaining intact in their graves with their skin smooth like the cheeks of a bride and their eyes alive and clear like emeralds.

2. See the graph "Growth in Genbank DNA Sequence Data" in chapter 10.
3. Cheng Zhang and Jianpeng Ma, "Enhanced Sampling and Applications in Protein Folding in Explicit Solvent," *Journal of Chemical Physics* 132, no. 24 (2010): 244101. See also http://folding.stanford.edu/English/About about the Folding@ home project, which has harnessed over five million computers around the world to simulate protein folding.
4. For a more complete description of this argument, see the section "[The Impact . . .] on the Intelligent Destiny of the Cosmos: Why We Are Probably Alone in the Universe" in chapter 6 of *The Singularity Is Near* by Ray Kurzweil (New York: Viking, 2005).
5. James D. Watson, *Discovering the Brain* (Washington, DC: National Academies Press, 1992).
6. Sebastian Seung, *Connectome: How the Brain's Wiring Makes Us Who We Are* (New York: Houghton Mifflin Harcourt, 2012).
7. "Mandelbrot Zoom," http://www.youtube.com/watch?v=gEw8xpb1aRA; "Fractal Zoom Mandelbrot Corner," http://www.youtube.com/watch?v=G _GBwuYuOOs.

Chapter 1: Thought Experiments on the World

1. Charles Darwin, *The Origin of Species* (P. F. Collier & Son, 1909), 185/95–96.
2. Darwin, *On the Origin of Species*, 751 (206.1.1-6), Peckham's Variorum edition, edited by Morse Peckham, *The Origin of Species by Charles Darwin: A Variorum Text* (Philadelphia: University of Pennsylvania Press, 1959).
3. R. Dahm, "Discovering DNA: Friedrich Miescher and the Early Years of Nucleic Acid Research," *Human Genetics* 122, no. 6 (2008): 565–81, doi:10.1007/s00439 -007-0433-0; PMID 17901982.

4. Valery N. Soyfer, "The Consequences of Political Dictatorship for Russian Science," *Nature Reviews Genetics* 2, no. 9 (2001): 723–29, doi:10.1038/35088598; PMID 11533721.

5. J. D. Watson and F. H. C. Crick, "A Structure for Deoxyribose Nucleic Acid," *Nature* 171 (1953): 737–38, http://www.nature.com/nature/dna50/watsoncrick .pdf and "Double Helix: 50 Years of DNA," *Nature* archive, http://www.nature.com /nature/dna50/archive.html.

6. Franklin died in 1958 and the Nobel Prize for the discovery of DNA was awarded in 1962. There is controversy as to whether or not she would have shared in that prize had she been alive in 1962.

7. Albert Einstein, "On the Electrodynamics of Moving Bodies" (1905). This paper established the special theory of relativity. See Robert Bruce Lindsay and Henry Margenau, *Foundations of Physics* (Woodbridge, CT: Ox Bow Press, 1981), 330.

8. "Crookes radiometer," Wikipedia, http://en.wikipedia.org/wiki/Crookes_radi ometer.

9. Note that some of the momentum of the photons is transferred to the air molecules in the bulb (since it is not a perfect vacuum) and then transferred from the heated air molecules to the vane.

10. Albert Einstein, "Does the Inertia of a Body Depend Upon Its Energy Content?" (1905). This paper established Einstein's famous formula $E = mc^2$.

11. "Albert Einstein's Letters to President Franklin Delano Roosevelt," http:// hypertextbook.com/eworld/einstein.shtml.

Chapter 3: A Model of the Neocortex:
The Pattern Recognition Theory of Mind

1. Some nonmammals, such as crows, parrots, and octopi, are reported to be capable of some level of reasoning; however, this is limited and has not been sufficient to create tools that have their own evolutionary course of development. These animals may have adapted other brain regions to perform a small number of levels of hierarchical thinking, but a neocortex is required for the relatively unrestricted hierarchical thinking that humans can perform.

2. V. B. Mountcastle, "An Organizing Principle for Cerebral Function: The Unit Model and the Distributed System" (1978), in Gerald M. Edelman and Vernon B. Mountcastle, *The Mindful Brain: Cortical Organization and the Group-Selective Theory of Higher Brain Function* (Cambridge, MA: MIT Press, 1982).

3. Herbert A. Simon, "The Organization of Complex Systems," in Howard H. Pattee, ed., *Hierarchy Theory: The Challenge of Complex Systems* (New York: George Braziller, Inc., 1973), http://blog.santafe.edu/wp-content/uploads/2009/03/simon 1973.pdf.

4. Marc D. Hauser, Noam Chomsky, and W. Tecumseh Fitch, "The Faculty of Language: What Is It, Who Has It, and How Did It Evolve?" *Science* 298 (November 2002): 1569–79, http://www.sciencemag.org/content/298/5598/1569.short.

5. The following passage from the book *Transcend: Nine Steps to Living Well Forever,* by Ray Kurzweil and Terry Grossman (New York: Rodale, 2009), describes this lucid dreaming technique in more detail:

> I've developed a method of solving problems while I sleep. I've perfected it for myself over several decades and have learned the subtle means by which this is likely to work better.
>
> I start out by assigning myself a problem when I get into bed. This can be any kind of problem. It could be a math problem, an issue with one of my inventions, a business strategy question, or even an interpersonal problem.
>
> I'll think about the problem for a few minutes, but I try not to solve it. That would just cut off the creative problem solving to come. I do try to think about it. What do I know about this? What form could a solution take? And then I go to sleep. Doing this primes my subconscious mind to work on the problem.
>
> Terry: Sigmund Freud pointed out that when we dream, many of the censors in our brain are relaxed, so that we might dream about things that are socially, culturally, or even sexually taboo. We can dream about weird things that we wouldn't allow ourselves to think about during the day. That's at least one reason why dreams are strange.
>
> Ray: There are also professional blinders that prevent people from thinking creatively, many of which come from our professional training, mental blocks such as "you can't solve a signal processing problem that way" or "linguistics is not supposed to use those rules." These mental assumptions are also relaxed in our dream state, so I'll dream about new ways of solving problems without being burdened by these daytime constraints.
>
> Terry: There's another part of our brain also not working when we dream, our rational faculties to evaluate whether an idea is reasonable. So that's another reason that weird or fantastic things happen in our dreams. When the elephant walks through the wall, we aren't shocked as to how the elephant could do this. We just say to our dream selves, "Okay, an elephant walked through the wall, no big deal." Indeed, if I wake up in the middle of the night, I often find that I've been dreaming in strange and oblique ways about the problem that I assigned myself.

Ray: The next step occurs in the morning in the halfway state between dreaming and being awake, which is often called *lucid dreaming*. In this state, I still have the feelings and imagery from my dreams, but now I do have my rational faculties. I realize, for example, that I am in a bed. And I could formulate the rational thought that I have a lot to do so I had better get out of bed. But that would be a mistake. Whenever I can, I will stay in bed and continue in this lucid dream state because that is key to this creative problem-solving method. By the way, this doesn't work if the alarm rings.

Reader: Sounds like the best of both worlds.

Ray: Exactly. I still have access to the dream thoughts about the problem I assigned myself the night before. But now I'm sufficiently conscious and rational to evaluate the new creative ideas that came to me during the night. I can determine which ones make sense. After perhaps 20 minutes of this, I invariably will have keen new insights into the problem.

I've come up with inventions this way (and spent the rest of the day writing a patent application), figured out how to organize material for a book such as this, and come up with useful ideas for a diverse set of problems. If I have a key decision to make, I will always go through this process, after which I am likely to have real confidence in my decision.

The key to the process is to let your mind go, to be nonjudgmental, and not to worry about how well the method is working. It is the opposite of a mental discipline. Think about the problem, but then let ideas wash over you as you fall asleep. Then in the morning, let your mind go again as you review the strange ideas that your dreams generated. I have found this to be an invaluable method for harnessing the natural creativity of my dreams.

Reader: Well, for the workaholics among us, we can now work in our dreams. Not sure my spouse is going to appreciate this.

Ray: Actually, you can think of it as getting your dreams to do your work for you.

Chapter 4: The Biological Neocortex

1. Steven Pinker, *How the Mind Works* (New York: Norton, 1997), 152–53.
2. D. O. Hebb, *The Organization of Behavior* (New York: John Wiley & Sons, 1949).
3. Henry Markram and Rodrigo Perrin, "Innate Neural Assemblies for Lego Memory," *Frontiers in Neural Circuits* 5, no. 6 (2011).

4. E-mail communication from Henry Markram, February 19, 2012.

5. Van J. Wedeen et al., "The Geometric Structure of the Brain Fiber Pathways," *Science* 335, no. 6076 (March 30, 2012).

6. Tai Sing Lee, "Computations in the Early Visual Cortex," *Journal of Physiology— Paris* 97 (2003): 121–39.

7. A list of papers can be found at http://cbcl.mit.edu/people/poggio/tpcv_short _pubs.pdf.

8. Daniel J. Felleman and David C. Van Essen, "Distributed Hierarchical Processing in the Primate Cerebral Cortex," *Cerebral Cortex* 1, no. 1 (January/February 1991): 1–47. A compelling analysis of the Bayesian mathematics of the top-down and bottom-up communication in the neocortex is provided by Tai Sing Lee in "Hierarchical Bayesian Inference in the Visual Cortex," *Journal of the Optical Society of America* 20, no. 7 (July 2003): 1434–48.

9. Uri Hasson et al., "A Hierarchy of Temporal Receptive Windows in Human Cortex," *Journal of Neuroscience* 28, no. 10 (March 5, 2008): 2539–50.

10. Marina Bedny et al., "Language Processing in the Occipital Cortex of Congenitally Blind Adults," *Proceedings of the National Academy of Sciences* 108, no. 11 (March 15, 2011): 4429–34.

11. Daniel E. Feldman, "Synaptic Mechanisms for Plasticity in Neocortex," *Annual Review of Neuroscience* 32 (2009): 33–55.

12. Aaron C. Koralek et al., "Corticostriatal Plasticity Is Necessary for Learning Intentional Neuroprosthetic Skills," *Nature* 483 (March 15, 2012): 331–35.

13. E-mail communication from Randal Koene, January 2012.

14. Min Fu, Xinzhu Yu, Ju Lu, and Yi Zuo, "Repetitive Motor Learning Induces Coordinated Formation of Clustered Dendritic Spines *in Vivo*," *Nature* 483 (March 1, 2012): 92–95.

15. Dario Bonanomi et al., "Ret Is a Multifunctional Coreceptor That Integrates Diffusible- and Contact-Axon Guidance Signals," *Cell* 148, no. 3 (February 2012): 568–82.

16. See endnote 7 in chapter 11.

Chapter 5: The Old Brain

1. Vernon B. Mountcastle, "The View from Within: Pathways to the Study of Perception," *Johns Hopkins Medical Journal* 136 (1975): 109–31.

2. B. Roska and F. Werblin, "Vertical Interactions Across Ten Parallel, Stacked Representations in the Mammalian Retina," *Nature* 410, no. 6828 (March 29, 2001): 583–87; "Eye Strips Images of All but Bare Essentials Before Sending Visual Information to Brain, UC Berkeley Research Shows," University of California at Berkeley news release, March 28, 2001, www.berkeley.edu/news/media/releases /2001/03/28_wers1.html.

3. Lloyd Watts, "Reverse-Engineering the Human Auditory Pathway," in J. Liu et al., eds., *WCCI 2012* (Berlin: Springer-Verlag, 2012), 47–59. Lloyd Watts, "Real-Time, High-Resolution Simulation of the Auditory Pathway, with Application to Cell-Phone Noise Reduction," *ISCAS* (June 2, 2010): 3821–24. For other papers see http://www.lloydwatts.com/publications.html.

4. See Sandra Blakeslee, "Humanity? Maybe It's All in the Wiring," *New York Times,* December 11, 2003, http://www.nytimes.com/2003/12/09/science/09BRAI .html.

5. T. E. J. Behrens et al., "Non-Invasive Mapping of Connections between Human Thalamus and Cortex Using Diffusion Imaging," *Nature Neuroscience* 6, no. 7 (July 2003): 750–57.

6. Timothy J. Buschman et al., "Neural Substrates of Cognitive Capacity Limitations," *Proceedings of the National Academy of Sciences* 108, no. 27 (July 5, 2011): 11252–55, http://www.pnas.org/content/108/27/11252.long.

7. Theodore W. Berger et al., "A Cortical Neural Prosthesis for Restoring and Enhancing Memory," *Journal of Neural Engineering* 8, no. 4 (August 2011).

8. Basis functions are nonlinear functions that can be combined linearly (by adding together multiple weighted-basis functions) to approximate any nonlinear function. A. Pouget and L. H. Snyder, "Computational Approaches to Sensorimotor Transformations," *Nature Neuroscience* 3, no. 11 Supplement (November 2000): 1192–98.

9. J. R. Bloedel, "Functional Heterogeneity with Structural Homogeneity: How Does the Cerebellum Operate?" *Behavioral and Brain Sciences* 15, no. 4 (1992): 666–78.

10. S. Grossberg and R. W. Paine, "A Neural Model of Cortico-Cerebellar Interactions during Attentive Imitation and Predictive Learning of Sequential Handwriting Movements," *Neural Networks* 13, no. 8–9 (October–November 2000): 999–1046.

11. Javier F. Medina and Michael D. Mauk, "Computer Simulation of Cerebellar Information Processing," *Nature Neuroscience* 3 (November 2000): 1205–11.

12. James Olds, "Pleasure Centers in the Brain," *Scientific American* (October 1956): 105–16. Aryeh Routtenberg, "The Reward System of the Brain," *Scientific American* 239 (November 1978): 154–64. K. C. Berridge and M. L. Kringelbach, "Affective Neuroscience of Pleasure: Reward in Humans and Other Animals," *Psychopharmacology* 199 (2008): 457–80. Morten L. Kringelbach, *The Pleasure Center: Trust Your Animal Instincts* (New York: Oxford University Press, 2009). Michael R. Liebowitz, *The Chemistry of Love* (Boston: Little, Brown, 1983). W. L. Witters and P. Jones-Witters, *Human Sexuality: A Biological Perspective* (New York: Van Nostrand, 1980).

Chapter 6: Transcendent Abilities

1. Michael Nielsen, *Reinventing Discovery: The New Era of Networked Science* (Princeton, NJ: Princeton University Press, 2012), 1–3. T. Gowers and M. Nielsen, "Massively Collaborative Mathematics," *Nature* 461, no. 7266 (2009): 879–81. "A Combinatorial Approach to Density Hales-Jewett," *Gowers's Weblog*, http://gow ers.wordpress.com/2009/02/01/a-combinatorial-approach-to-density-hales-jewett/. Michael Nielsen, "The Polymath Project: Scope of Participation," March 20, 2009, http://michaelnielsen.org/blog/?p=584. Julie Rehmeyer, "SIAM: Massively Collaborative Mathematics," Society for Industrial and Applied Mathematics, April 1, 2010, http://www.siam.org/news/news.php?id=1731.
2. P. Dayan and Q. J. M. Huys, "Serotonin, Inhibition, and Negative Mood," *PLoS Computational Biology* 4, no. 1 (2008), http://compbiol.plosjournals.org /perlserv/?request=get-document&doi=10.1371/journal.pcbi.0040004.

Chapter 7: The Biologically Inspired Digital Neocortex

1. Gary Cziko, *Without Miracles: Universal Selection Theory and the Second Darwinian Revolution* (Cambridge, MA: MIT Press, 1955).
2. David Dalrymple has been a mentee of mine since he was eight years old in 1999. You can read his background here: http://esp.mit.edu/learn/teachers/davidad/bio .html, and http://www.brainsciences.org/Research-Team/mr-david-dalrymple .html.
3. Jonathan Fildes, "Artificial Brain '10 Years Away,'" BBC News, July 22, 2009, http://news.bbc.co.uk/2/hi/8164060.stm. See also the video "Henry Markram on Simulating the Brain: The Next Decisive Years," http://www.kurzweilai.net /henry-markram-simulating-the-brain-next-decisive-years.
4. M. Mitchell Waldrop, "Computer Modelling: Brain in a Box," *Nature News*, February 22, 2012, http://www.nature.com/news/computer-modelling-brain-in-a-box-1 .10066.
5. Jonah Lehrer, "Can a Thinking, Remembering, Decision-Making Biologically Accurate Brain Be Built from a Supercomputer?" *Seed*, http://seedmagazine .com/content/article/out_of_the_blue/.
6. Fildes, "Artificial Brain '10 Years Away.'"
7. See http://www.humanconnectomeproject.org/.
8. Anders Sandberg and Nick Bostrom, *Whole Brain Emulation: A Roadmap,* Technical Report #2008–3 (2008), Future of Humanity Institute, Oxford University, www.fhi.ox.ac.uk/reports/2008-3.pdf.
9. Here is the basic schema for a neural net algorithm. Many variations are possible, and the designer of the system needs to provide certain critical parameters and methods, detailed on the following pages.

Creating a neural net solution to a problem involves the following steps:

Define the input.

Define the topology of the neural net (i.e., the layers of neurons and the connections between the neurons).

Train the neural net on examples of the problem.

Run the trained neural net to solve new examples of the problem.

Take your neural net company public.

These steps (except for the last one) are detailed below:

The Problem Input

The problem input to the neural net consists of a series of numbers. This input can be:

In a visual pattern recognition system, a two-dimensional array of numbers representing the pixels of an image; or

In an auditory (e.g., speech) recognition system, a two-dimensional array of numbers representing a sound, in which the first dimension represents parameters of the sound (e.g., frequency components) and the second dimension represents different points in time; or

In an arbitrary pattern recognition system, an n-dimensional array of numbers representing the input pattern.

Defining the Topology

To set up the neural net, the architecture of each neuron consists of:

Multiple inputs in which each input is "connected" to either the output of another neuron or one of the input numbers.

Generally, a single output, which is connected to either the input of another neuron (which is usually in a higher layer) or the final output.

Set Up the First Layer of Neurons

Create N_0 neurons in the first layer. For each of these neurons, "connect" each of the multiple inputs of the neuron to "points" (i.e., numbers) in the problem input. These connections can be determined randomly or using an evolutionary algorithm (see below).

Assign an initial "synaptic strength" to each connection created. These weights can start out all the same, can be assigned randomly, or can be determined in another way (see below).

Set Up the Additional Layers of Neurons

Set up a total of M layers of neurons. For each layer, set up the neurons in that layer. For $layer_i$:

Create N_i neurons in $layer_i$. For each of these neurons, "connect" each of the multiple inputs of the neuron to the outputs of the neurons in $layer_{i-1}$ (see variations below).

Assign an initial "synaptic strength" to each connection created. These weights can start out all the same, can be assigned randomly, or can be determined in another way (see below).

The outputs of the neurons in $layer_M$ are the outputs of the neural net (see variations below).

The Recognition Trials

How Each Neuron Works

Once the neuron is set up, it does the following for each recognition trial:

Each weighted input to the neuron is computed by multiplying the output of the other neuron (or initial input) that the input to this neuron is connected to by the synaptic strength of that connection.

All of these weighted inputs to the neuron are summed.

If this sum is greater than the firing threshold of this neuron, then this neuron is considered to fire and its output is 1. Otherwise, its output is 0 (see variations below).

Do the Following for Each Recognition Trial

For each layer, from $layer_0$ to $layer_M$:
For each neuron in the layer:

Sum its weighted inputs (each weighted input = the output of the other neuron [or initial input] that the input to this neuron is connected to, multiplied by the synaptic strength of that connection).

If this sum of weighted inputs is greater than the firing threshold for this neuron, set the output of this neuron = 1, otherwise set it to 0.

To Train the Neural Net

Run repeated recognition trials on sample problems.

After each trial, adjust the synaptic strengths of all the interneuronal connections to improve the performance of the neural net on this trial (see the discussion below on how to do this).

Continue this training until the accuracy rate of the neural net is no longer improving (i.e., reaches an asymptote).

Key Design Decisions

In the simple schema above, the designer of this neural net algorithm needs to determine at the outset:

What the input numbers represent.

The number of layers of neurons.

The number of neurons in each layer. (Each layer does not necessarily need to have the same number of neurons.)

The number of inputs to each neuron in each layer. The number of inputs (i.e., interneuronal connections) can also vary from neuron to neuron and from layer to layer.

The actual "wiring" (i.e., the connections). For each neuron in each layer, this consists of a list of other neurons, the outputs of which constitute the inputs to this neuron. This represents a key design area. There are a number of possible ways to do this:

(1) Wire the neural net randomly; or
(2) Use an evolutionary algorithm (see below) to determine an optimal wiring; or
(3) Use the system designer's best judgment in determining the wiring.

The initial synaptic strengths (i.e., weights) of each connection. There are a number of possible ways to do this:

(1) Set the synaptic strengths to the same value; or
(2) Set the synaptic strengths to different random values; or
(3) Use an evolutionary algorithm to determine an optimal set of initial values; or
(4) Use the system designer's best judgment in determining the initial values.

The firing threshold of each neuron.

Determine the output. The output can be:

(1) the outputs of layer$_M$ of neurons; or

(2) the output of a single output neuron, the inputs of which are the outputs of the neurons in layer$_M$; or

(3) a function of (e.g., a sum of) the outputs of the neurons in layer$_M$; or

(4) another function of neuron outputs in multiple layers.

Determine how the synaptic strengths of all the connections are adjusted during the training of this neural net. This is a key design decision and is the subject of a great deal of research and discussion. There are a number of possible ways to do this:

(1) For each recognition trial, increment or decrement each synaptic strength by a (generally small) fixed amount so that the neural net's output more closely matches the correct answer. One way to do this is to try both incrementing and decrementing and see which has the more desirable effect. This can be time-consuming, so other methods exist for making local decisions on whether to increment or decrement each synaptic strength.

(2) Other statistical methods exist for modifying the synaptic strengths after each recognition trial so that the performance of the neural net on that trial more closely matches the correct answer.

Note that neural net training will work even if the answers to the training trials are not all correct. This allows using real-world training data that may have an inherent error rate. One key to the success of a neural net–based recognition system is the amount of data used for training. Usually a very substantial amount is needed to obtain satisfactory results. As with human students, the amount of time that a neural net spends learning its lessons is a key factor in its performance.

Variations

Many variations of the above are feasible. For example:

There are different ways of determining the topology. In particular, the interneuronal wiring can be set either randomly or using an evolutionary algorithm.

There are different ways of setting the initial synaptic strengths.

The inputs to the neurons in layer$_i$ do not necessarily need to come from the outputs of the neurons in layer$_{i-1}$. Alternatively, the inputs to the neurons in each layer can come from any lower layer or any layer.

There are different ways to determine the final output.

The method described above results in an "all or nothing" (1 or 0) firing called a nonlinearity. There are other nonlinear functions that can be used. Commonly a function is used that goes from 0 to 1 in a rapid but more gradual fashion. Also, the outputs can be numbers other than 0 and 1.

The different methods for adjusting the synaptic strengths during training represent key design decisions.

The above schema describes a "synchronous" neural net, in which each recognition trial proceeds by computing the outputs of each layer, starting with layer$_0$ through layer$_M$. In a true parallel system, in which each neuron is operating independently of the others, the neurons can operate "asynchronously" (i.e., independently). In an asynchronous approach, each neuron is constantly scanning its inputs and fires whenever the sum of its weighted inputs exceeds its threshold (or whatever its output function specifies).

10. Robert Mannell, "Acoustic Representations of Speech," 2008, http://clas.mq.edu .au/acoustics/frequency/acoustic_speech.html.
11. Here is the basic schema for a genetic (evolutionary) algorithm. Many variations are possible, and the designer of the system needs to provide certain critical parameters and methods, detailed below.

The Evolutionary Algorithm

Create N solution "creatures." Each one has:

A genetic code: a sequence of numbers that characterize a possible solution to the problem. The numbers can represent critical parameters, steps to a solution, rules, etc.

For each generation of evolution, do the following:

Do the following for each of the N solution creatures:

Apply this solution creature's solution (as represented by its genetic code) to the problem, or simulated environment. Rate the solution.

Pick the L solution creatures with the highest ratings to survive into the next generation.

Eliminate the $(N - L)$ nonsurviving solution creatures.

Create $(N - L)$ new solution creatures from the L surviving solution creatures by:

(1) Making copies of the L surviving creatures. Introduce small random variations into each copy; or
(2) Create additional solution creatures by combining parts of the genetic code (using "sexual" reproduction, or otherwise combining portions of the chromosomes) from the L surviving creatures; or
(3) Do a combination of (1) and (2).

Determine whether or not to continue evolving:

Improvement = (highest rating in this generation) – (highest rating in the previous generation).

If Improvement < Improvement Threshold then we're done.

The solution creature with the highest rating from the last generation of evolution has the best solution. Apply the solution defined by its genetic code to the problem.

Key Design Decisions

In the simple schema above, the designer needs to determine at the outset:

Key parameters:

N

L

Improvement threshold.

What the numbers in the genetic code represent and how the solution is computed from the genetic code.

A method for determining the N solution creatures in the first generation. In general, these need only be "reasonable" attempts at a solution. If these first-generation solutions are too far afield, the evolutionary algorithm may have difficulty converging on a good solution. It is often worthwhile to create the initial solution creatures in such a way that they are reasonably diverse. This will help prevent the evolutionary process from just finding a "locally" optimal solution.

How the solutions are rated.

How the surviving solution creatures reproduce.

Variations

Many variations of the above are feasible. For example:

There does not need to be a fixed number of surviving solution creatures (L) from each generation. The survival rule(s) can allow for a variable number of survivors.

There does not need to be a fixed number of new solution creatures created in each generation ($N - L$). The procreation rules can be independent of the size of the population. Procreation can be related to survival, thereby allowing the fittest solution creatures to procreate the most.

The decision as to whether or not to continue evolving can be varied. It can consider more than just the highest-rated solution creature from the most recent generation(s). It can also consider a trend that goes beyond just the last two generations.

12. Dileep George, "How the Brain Might Work: A Hierarchical and Temporal Model for Learning and Recognition" (PhD dissertation, Stanford University, June 2008).

13. A. M. Turing, "Computing Machinery and Intelligence," *Mind*, October 1950.

14. Hugh Loebner has a "Loebner Prize" competition that is run each year. The Loebner silver medal will go to a computer that passes Turing's original text-only test. The gold medal will go to a computer that can pass a version of the test that includes audio and video input and output. In my view, the inclusion of audio and video does not actually make the test more challenging.

15. "Cognitive Assistant That Learns and Organizes," Artificial Intelligence Center, SRI International, http://www.ai.sri.com/project/CALO.

16. Dragon Go! Nuance Communications, Inc., http://www.nuance.com/products /dragon-go-in-action/index.htm.

17. "Overcoming Artificial Stupidity," *WolframAlpha Blog*, April 17, 2012, http:// blog.wolframalpha.com/author/stephenwolfram/.

Chapter 8: The Mind as Computer

1. Salomon Bochner, *A Biographical Memoir of John von Neumann* (Washington, DC: National Academy of Sciences, 1958).

2. A. M. Turing, "On Computable Numbers, with an Application to the Entscheidungsproblem," *Proceedings of the London Mathematical Society* Series 2, vol. 42 (1936–37): 230–65, http://www.comlab.ox.ac.uk/activities/ieg/e-library/sources /tp2-ie.pdf. A. M. Turing, "On Computable Numbers, with an Application to the Entscheidungsproblem: A Correction," *Proceedings of the London Mathematical Society* 43 (1938): 544–46.

3. John von Neumann, "First Draft of a Report on the EDVAC," Moore School of Electrical Engineering, University of Pennsylvania, June 30, 1945. John von Neumann, "A Mathematical Theory of Communication," *Bell System Technical Journal*, July and October 1948.

4. Jeremy Bernstein, *The Analytical Engine: Computers—Past, Present, and Future*, rev. ed. (New York: William Morrow & Co., 1981).

5. "Japan's K Computer Tops 10 Petaflop/s to Stay Atop TOP500 List," *Top 500*, November 11, 2011, http://top500.org/lists/2011/11/press-release.

6. Carver Mead, *Analog VLSI and Neural Systems* (Reading, MA: Addison-Wesley, 1986).

7. "IBM Unveils Cognitive Computing Chips," IBM news release, August 18, 2011, http://www-03.ibm.com/press/us/en/pressrelease/35251.wss.

8. "Japan's K Computer Tops 10 Petaflop/s to Stay Atop TOP500 List."

Chapter 9: Thought Experiments on the Mind

1. John R. Searle, "I Married a Computer," in Jay W. Richards, ed., *Are We Spiritual Machines? Ray Kurzweil vs. the Critics of Strong AI* (Seattle: Discovery Institute, 2002).

2. Stuart Hameroff, *Ultimate Computing: Biomolecular Consciousness and Nanotechnology* (Amsterdam: Elsevier Science, 1987).

3. P. S. Sebel et al., "The Incidence of Awareness during Anesthesia: A Multicenter United States Study," *Anesthesia and Analgesia* 99 (2004): 833–39.

4. Stuart Sutherland, *The International Dictionary of Psychology* (New York: Macmillan, 1990).

5. David Cockburn, "Human Beings and Giant Squids," *Philosophy* 69, no. 268 (April 1994): 135–50.

6. Ivan Petrovich Pavlov, from a lecture given in 1913, published in *Lectures on Conditioned Reflexes: Twenty-Five Years of Objective Study of the Higher Nervous Activity [Behavior] of Animals* (London: Martin Lawrence, 1928), 222.

7. Roger W. Sperry, from James Arthur Lecture on the Evolution of the Human Brain, 1964, p. 2.

8. Henry Maudsley, "The Double Brain," *Mind* 14, no. 54 (1889): 161–87.

9. Susan Curtiss and Stella de Bode, "Language after Hemispherectomy," *Brain and Cognition* 43, nos. 1–3 (June–August 2000): 135–38.

10. E. P. Vining et al., "Why Would You Remove Half a Brain? The Outcome of 58 Children after Hemispherectomy—the Johns Hopkins Experience: 1968 to 1996," Pediatrics 100 (August 1997): 163–71. M. B. Pulsifer et al., "The Cognitive Outcome of Hemispherectomy in 71 Children," *Epilepsia* 45, no. 3 (March 2004): 243–54.

11. S. McClelland III and R. E. Maxwell, "Hemispherectomy for Intractable Epilepsy in Adults: The First Reported Series," *Annals of Neurology* 61, no. 4 (April 2007): 372–76.

12. Lars Muckli, Marcus J. Naumerd, and Wolf Singer, "Bilateral Visual Field Maps in a Patient with Only One Hemisphere," *Proceedings of the National Academy of Sciences* 106, no. 31 (August 4, 2009), http://dx.doi.org/10.1073/pnas.0809688106.

13. Marvin Minsky, *The Society of Mind* (New York: Simon and Schuster, 1988).

14. F. Fay Evans-Martin, *The Nervous System* (New York: Chelsea House, 2005), http://www.scribd.com/doc/5012597/The-Nervous-System.

15. Benjamin Libet, *Mind Time: The Temporal Factor in Consciousness* (Cambridge, MA: Harvard University Press, 2005).

16. Daniel C. Dennett, *Freedom Evolves* (New York: Viking, 2003).

17. Michael S. Gazzaniga, *Who's in Charge? Free Will and the Science of the Brain* (New York: Ecco/HarperCollins, 2011).
18. David Hume, *An Enquiry Concerning Human Understanding* (1765), 2nd ed., edited by Eric Steinberg (Indianapolis: Hackett, 1993).
19. Arthur Schopenhauer, *The Wisdom of Life*.
20. Arthur Schopenhauer, *On the Freedom of the Will* (1839).
21. From Raymond Smullyan, *5000 B.C. and Other Philosophical Fantasies* (New York: St. Martin's Press, 1983).
22. For an insightful and entertaining examination of similar issues of identity and consciousness, see Martine Rothblatt, "The Terasem Mind Uploading Experiment," *International Journal of Machine Consciousness* 4, no. 1 (2012): 141–58. In this paper, Rothblatt examines the issue of identity with regard to software that emulates a person based on "a database of video interviews and associated information about a predecessor person." In this proposed future experiment, the software is successfully emulating the person it is based on.
23. "How Do You Persist When Your Molecules Don't?" *Science and Consciousness Review* 1, no. 1 (June 2004), http://www.sci-con.org/articles/20040601.html.

Chapter 10: The Law of Accelerating Returns Applied to the Brain

1. "DNA Sequencing Costs," National Human Genome Research Institute, NIH, http://www.genome.gov/sequencingcosts/.
2. "Genetic Sequence Data Bank, Distribution Release Notes," December 15, 2009, National Center for Biotechnology Information, National Library of Medicine, ftp://ftp.ncbi.nih.gov/genbank/gbrel.txt.
3. "DNA Sequencing—The History of DNA Sequencing," January 2, 2012, http://www.dnasequencing.org/history-of-dna.
4. "Cooper's Law," ArrayComm, http://www.arraycomm.com/technology/coopers-law.
5. "The Zettabyte Era," Cisco, http://www.cisco.com/en/US/solutions/collateral/ns341/ns525/ns537/ns705/ns827/VNI_Hyperconnectivity_WP.html, and "Number of Internet Hosts," Internet Systems Consortium, http://www.isc.org/solutions/survey/history.
6. TeleGeography © PriMetrica, Inc., 2012.
7. Dave Kristula, "The History of the Internet" (March 1997, update August 2001), http://www.davesite.com/webstation/net-history.shtml; Robert Zakon, "Hobbes' Internet Timeline v8.0," http://www.zakon.org/robert/internet/timeline; Quest Communications, 8-K for 9/13/1998 EX-99.1; *Converge! Network Digest,* December 5, 2002, http://www.convergedigest.com/Daily/daily.asp?vn=v9n229&fecha=December%2005,%202002; Jim Duffy, "AT&T Plans Backbone Upgrade to 40G," *Computerworld,* June 7, 2006, http://www.computerworld.com/action/article

.do?command=viewArticleBasic&articleId=9001032; "40G: The Fastest Connection You Can Get?" InternetNews.com, November 2, 2007, http://www.internetnews .com/infra/article.php/3708936; "Verizon First Global Service Provider to Deploy 100G on U.S. Long-Haul Network," news release, Verizon, http://newscenter .verizon.com/press-releases/verizon/2011/verizon-first-global-service.html.

8. Facebook, "Key Facts," http://newsroom.fb.com/content/default.aspx?NewsArea Id=22.

9. http://www.kurzweilai.net/how-my-predictions-are-faring.

10. **Calculations per Second per $1,000**

Year	Calculations per Second per $1,000	Machine	Natural Logarithm (calcs/sec/$k)
1900	5.82E−06	Analytical Engine	−12.05404
1908	1.30E−04	Hollerith Tabulator	−8.948746
1911	5.79E−05	Monroe Calculator	−9.757311
1919	1.06E−03	IBM Tabulator	−6.84572
1928	6.99E−04	National Ellis 3000	−7.265431
1939	8.55E−03	Zuse 2	−4.762175
1940	1.43E−02	Bell Calculator Model 1	−4.246797
1941	4.63E−02	Zuse 3	−3.072613
1943	5.31E+00	Colossus	1.6692151
1946	7.98E−01	ENIAC	−0.225521
1948	3.70E−01	IBM SSEC	−0.994793
1949	1.84E+00	BINAC	0.6081338
1949	1.04E+00	EDSAC	0.0430595
1951	1.43E+00	Univac I	0.3576744
1953	6.10E+00	Univac 1103	1.8089443
1953	1.19E+01	IBM 701	2.4748563
1954	3.67E−01	EDVAC	−1.002666
1955	1.65E+01	Whirlwind	2.8003255
1955	3.44E+00	IBM 704	1.2348899
1958	3.26E−01	Datamatic 1000	−1.121779
1958	9.14E−01	Univac II	−0.089487

Year	Calculations per Second per $1,000	Machine	Natural Logarithm (calcs/sec/$k)
1960	1.51E+00	IBM 1620	0.4147552
1960	1.52E+02	DEC PDP-1	5.0205856
1961	2.83E+02	DEC PDP-4	5.6436786
1962	2.94E+01	Univac III	3.3820146
1964	1.59E+02	CDC 6600	5.0663853
1965	4.83E+02	IBM 1130	6.1791882
1965	1.79E+03	DEC PDP-8	7.4910876
1966	4.97E+01	IBM 360 Model 75	3.9064073
1968	2.14E+02	DEC PDP-10	5.3641051
1973	7.29E+02	Intellec-8	6.5911249
1973	3.40E+03	Data General Nova	8.1318248
1975	1.06E+04	Altair 8800	9.2667207
1976	7.77E+02	DEC PDP-11 Model 70	6.6554404
1977	3.72E+03	Cray 1	8.2214789
1977	2.69E+04	Apple II	10.198766
1979	1.11E+03	DEC VAX 11 Model 780	7.0157124
1980	5.62E+03	Sun-1	8.6342649
1982	1.27E+05	IBM PC	11.748788
1982	1.27E+05	Compaq Portable	11.748788
1983	8.63E+04	IBM AT-80286	11.365353
1984	8.50E+04	Apple Macintosh	11.350759
1986	5.38E+05	Compaq Deskpro 386	13.195986
1987	2.33E+05	Apple Mac II	12.357076
1993	3.55E+06	Pentium PC	15.082176
1996	4.81E+07	Pentium PC	17.688377
1998	1.33E+08	Pentium II PC	18.708113
1999	7.03E+08	Pentium III PC	20.370867
2000	1.09E+08	IBM ASCI White	18.506858

Year	Calculations per Second per $1,000	Machine	Natural Logarithm (calcs/sec/$k)
2000	3.40E+08	Power Macintosh G4/500	19.644456
2003	2.07E+09	Power Macintosh G5 2.0	21.450814
2004	3.49E+09	Dell Dimension 8400	21.973168
2005	6.36E+09	Power Mac G5 Quad	22.573294
2008	3.50E+10	Dell XPS 630	24.278614
2008	2.07E+10	Mac Pro	23.7534
2009	1.63E+10	Intel Core i7 Desktop	23.514431
2010	5.32E+10	Intel Core i7 Desktop	24.697324

11. Top 500 Supercomputer Sites, http://top500.org/.

12. "Microprocessor Quick Reference Guide," Intel Research, http://www.intel.com/pressroom/kits/quickreffam.htm.

13. 1971–2000: VLSI Research Inc.

2001–2006: *The International Technology Roadmap for Semiconductors,* 2002 Update and 2004 Update, Table 7a, "Cost—Near-term Years," "DRAM cost/bit at (packaged microcents) at production."

2007–2008: *The International Technology Roadmap for Semiconductors,* 2007, Tables 7a and 7b, "Cost—Near-term Years," "Cost—Long-term Years," http://www.itrs.net/Links/2007ITRS/ExecSum2007.pdf.

2009–2022: *The International Technology Roadmap for Semiconductors,* 2009, Tables 7a and 7b, "Cost—Near-term Years," "Cost—Long-term Years," http://www.itrs.net/Links/2009ITRS/Home2009.htm.

14. To make all dollar values comparable, computer prices for all years were converted to their year 2000 dollar equivalent using the Federal Reserve Board's CPI data at http://minneapolisfed.org/research/data/us/calc/. For example, $1 million in 1960 is equivalent to $5.8 million in 2000, and $1 million in 2004 is equivalent to $0.91 million in 2000.

1949: http://www.cl.cam.ac.uk/UoCCL/misc/EDSAC99/statistics.html, http://www.davros.org/misc/chronology.html.

1951: Richard E. Matick, *Computer Storage Systems and Technology* (New York: John Wiley & Sons, 1977); http://inventors.about.com/library/weekly/aa062398.htm.

1955: Matick, *Computer Storage Systems and Technology;* OECD, 1968, http://members.iinet.net.au/~dgreen/timeline.html.

1960:   ftp://rtfm.mit.edu/pub/usenet/alt.sys.pdp8/PDP-8_Frequently_Asked
Questions%28posted_every_other_month%29; http://www.dbit.com/~greeng3
/pdp1/pdp1.html#INTRODUCTION.

1962:   ftp://rtfm.mit.edu/pub/usenet/alt.sys.pdp8/PDP-8_Frequently_Asked
Questions%28posted_every_other_month%29.

1964: Matick, *Computer Storage Systems and Technology*; http://www.research
.microsoft.com/users/gbell/craytalk; http://www.ddj.com/documents/s=1493/ddj
0005hc/.

1965: Matick, *Computer Storage Systems and Technology*; http://www.fourmilab
.ch/documents/univac/config1108.html; http://www.frobenius.com/univac.htm.

1968: Data General.

1969, 1970: http://www.eetimes.com/special/special_issues/millennium/mile
stones/whittier.html.

1974: Scientific Electronic Biological Computer Consulting (SCELBI).

1975–1996: *Byte* magazine advertisements.

1997–2000: *PC Computing* magazine advertisements.

2001: www.pricewatch.com (http://www.jc-news.com/parse.cgi?news/price
watch/raw/pw-010702).

2002: www.pricewatch.com (http://www.jc-news.com/parse.cgi?news/price
watch/raw/pw-020624).

2003: http://sharkyextreme.com/guides/WMPG/article.php/10706_2227191_2.

2004: http://www.pricewatch.com (11/17/04).

2008: http://www.pricewatch.com (10/02/08) ($16.61).

15. Dataquest/Intel and Pathfinder Research:

Year	$	Log ($)
1968	1.00000000	0
1969	0.85000000	−0.16252
1970	0.60000000	−0.51083
1971	0.30000000	−1.20397
1972	0.15000000	−1.89712
1973	0.10000000	−2.30259
1974	0.07000000	−2.65926
1975	0.02800000	−3.57555
1976	0.01500000	−4.19971
1977	0.00800000	−4.82831

Year	$	Log ($)
1978	0.00500000	−5.29832
1979	0.00200000	−6.21461
1980	0.00130000	−6.64539
1981	0.00082000	−7.10621
1982	0.00040000	−7.82405
1983	0.00032000	−8.04719
1984	0.00032000	−8.04719
1985	0.00015000	−8.80488
1986	0.00009000	−9.31570
1987	0.00008100	−9.42106
1988	0.00006000	−9.72117
1989	0.00003500	−10.2602
1990	0.00002000	−10.8198
1991	0.00001700	−10.9823
1992	0.00001000	−11.5129
1993	0.00000900	−11.6183
1994	0.00000800	−11.7361
1995	0.00000700	−11.8696
1996	0.00000500	−12.2061
1997	0.00000300	−12.7169
1998	0.00000140	−13.4790
1999	0.00000095	−13.8668
2000	0.00000080	−14.0387
2001	0.00000035	−14.8653
2002	0.00000026	−15.1626
2003	0.00000017	−15.5875
2004	0.00000012	−15.9358
2005	0.000000081	−16.3288

Year	$	Log ($)
2006	0.000000063	−16.5801
2007	0.000000024	−17.5452
2008	0.000000016	−17.9507

16. Steve Cullen, In-Stat, September 2008, www.instat.com.

Year	Mbits	Bits
1971	921.6	9.216E+08
1972	3788.8	3.789E+09
1973	8294.4	8.294E+09
1974	19865.6	1.987E+10
1975	42700.8	4.270E+10
1976	130662.4	1.307E+11
1977	276070.4	2.761E+11
1978	663859.2	6.639E+11
1979	1438720.0	1.439E+12
1980	3172761.6	3.173E+12
1981	4512665.6	4.513E+12
1982	11520409.6	1.152E+13
1983	29648486.4	2.965E+13
1984	68418764.8	6.842E+13
1985	87518412.8	8.752E+13
1986	192407142.4	1.924E+14
1987	255608422.4	2.556E+14
1988	429404979.2	4.294E+14
1989	631957094.4	6.320E+14
1990	950593126.4	9.506E+14
1991	1546590618	1.547E+15
1992	2845638656	2.846E+15

Year	Mbits	Bits
1993	4177959322	4.178E+15
1994	7510805709	7.511E+15
1995	13010599936	1.301E+16
1996	23359078007	2.336E+16
1997	45653879161	4.565E+16
1998	85176878105	8.518E+16
1999	1.47327E+11	1.473E+17
2000	2.63636E+11	2.636E+17
2001	4.19672E+11	4.197E+17
2002	5.90009E+11	5.900E+17
2003	8.23015E+11	8.230E+17
2004	1.32133E+12	1.321E+18
2005	1.9946E+12	1.995E+18
2006	2.94507E+12	2.945E+18
2007	5.62814E+12	5.628E+18

17. "Historical Notes about the Cost of Hard Drive Storage Space," http://www.littletechshoppe.com/ns1625/winchest.html; *Byte* magazine advertisements, 1977–1998; *PC Computing* magazine advertisements, 3/1999; *Understanding Computers: Memory and Storage* (New York: Time Life, 1990); http://www.cedmagic.com/history/ibm-305-ramac.html; John C. McCallum, "Disk Drive Prices (1955–2012)," http://www.jcmit.com/diskprice.htm; IBM, "Frequently Asked Questions," http://www-03.ibm.com/ibm/history/documents/pdf/faq.pdf; IBM, "IBM 355 Disk Storage Unit," http://www-03.ibm.com/ibm/history/exhibits/storage/storage_355.html; IBM, "IBM 3380 Direct Access Storage Device," http://www.03-ibm.com/ibm/history/exhibits/storage/storage_3380.html.

18. "Without Driver or Map, Vans Go from Italy to China," *Sydney Morning Herald*, October 29, 2010, http://www.smh.com.au/technology/technology-news/without-driver-or-map-vans-go-from-italy-to-china-20101029-176ja.html.

19. KurzweilAI.net.

20. Adapted with permission from Amiram Grinvald and Rina Hildesheim, "VSDI: A New Era in Functional Imaging of Cortical Dynamics," *Nature Reviews Neuroscience* 5 (November 2004): 874–85.

The main tools for imaging the brain are shown in this diagram. Their capabilities are depicted by the shaded rectangles.

Spatial resolution refers to the smallest dimension that can be measured with a technique. Temporal resolution is imaging time or duration. There are trade-offs with each technique. For example, EEG (electroencephalography), which measures "brain waves" (electrical signals from neurons), can measure very rapid brain waves (occurring in short time intervals), but can only sense signals near the surface of the brain.

In contrast, fMRI (functional magnetic resonance imaging), which uses a special MRI machine to measure blood flow to neurons (indicating neuron activity), can sense a lot deeper in the brain (and spinal cord) and with higher resolution, down to tens of microns (millionths of a meter). However, fMRI operates very slowly compared with EEG.

These are noninvasive techniques (no surgery or drugs are required). MEG (magnetoencephalography) is another noninvasive technique. It detects magnetic fields generated by neurons. MEG and EEG can resolve events with a temporal resolution of down to 1 millisecond, but better than fMRI, which can at best resolve events with a resolution of several hundred milliseconds. MEG also accurately pinpoints sources in primary auditory, somatosensory, and motor areas.

Optical imaging covers almost the entire range of spatial and temporal resolutions, but is invasive. VSDI (voltage-sensitive dyes) is the most sensitive method of measuring brain activity, but is limited to measurements near the surface of the cortex of animals.

The exposed cortex is covered with a transparent sealed chamber; after the cortex is stained with a suitable voltage-sensitive dye, it is illuminated with light and a sequence of images is taken with a high-speed camera. Other optical techniques used in the lab include ion imaging (typically calcium or sodium ions) and fluorescence imaging systems (confocal imaging and multiphoton imaging).

Other lab techniques include PET (positron emission tomography, a nuclear medicine imaging technique that produces a 3-D image), 2DG (2-deoxyglucose postmortem histology, or tissue analysis), lesions (involves damaging neurons in an animal and observing the effects), patch clamping (to measure ion currents across biological membranes), and electron microscopy (using an electron beam to examine tissues or cells at a very fine scale). These techniques can also be integrated with optical imaging.

21. MRI spatial resolution in microns (μm), 1980–2012:

Year	Resolution in microns	Citation	URL
2012	125	"Characterization of Cerebral White Matter Properties Using Quantitative Magnetic Resonance Imaging Stains"	http://dx.doi.org/10.1089 /brain.2011.0071
2010	200	"Study of Brain Anatomy with High-Field MRI: Recent Progress"	http://dx.doi .org/10.1016/j .mri.2010.02.007
2010	250	"High-Resolution Phased-Array MRI of the Human Brain at 7 Tesla: Initial Experience in Multiple Sclerosis Patients"	http://dx.doi.org/10.1111 /j.1552-6569.2008.00338.x
1994	1,000	"Mapping Human Brain Activity in Vivo"	http://www.ncbi.nlm .nih.gov/pmc/articles /PMC1011409/
1989	1,700	"Neuroimaging in Patients with Seizures of Probable Frontal Lobe Origin"	http://dx.doi .org/10.1111/j.1528-1157 .1989.tb05470.x
1985	1,700	"A Study of the Septum Pellucidum and Corpus Callosum in Schizophrenia with MR Imaging"	http://dx.doi .org/10.1111/j.1600-0447 .1985.tb02634.x
1983	1,700	"Clinical Efficiency of Nuclear Magnetic Resonance Imaging"	http://radiology.rsna.org /content/146/1/123.short
1980	5,000	"In Vivo NMR Imaging in Medicine: The Aberdeen Approach, Both Physical and Biological [and Discussion]"	http://dx.doi.org/10.1098 /rstb.1980.0071

22. Spatial resolution in nanometers (nm) of destructive imaging techniques, 1983–2011:

Year	x-y res (nm)	Citation	URL	Technique	Notes
2011	4	"Focused Ion Beam Milling and Scanning Electron Microscopy of Brain Tissue"	http://dx.doi.org/10.3791/2588	Focused ion beam/scanning electron microscope (FIB/SEM)	
2011	4	"Volume Electron Microscopy for Neuronal Circuit Reconstruction"	http://dx.doi.org/10.1016/j.conb.2011.10.022	Scanning electron microscopy (SEM)	
2011	4	"Volume Electron Microscopy for Neuronal Circuit Reconstruction"	http://dx.doi.org/10.1016/j.conb.2011.10.022	Transmission electron microscopy (TEM)	
2004	13	"Serial Block-Face Scanning Electron Microscopy to Reconstruct Three-Dimensional Tissue Nanostructure"	http://dx.doi.org/10.1371/journal.pbio.0020329	Serial block-face scanning electron microscopy (SBF-SEM)	Result quoted in http://faculty.cs.tamu.edu/choe/ftp/publications/choe.hpc08-preprint.pdf, provided by Yoonsuck Choe.

Year	x-y res (nm)	Citation	URL	Technique	Notes
2004	20	"Wet SEM: A Novel Method for Rapid Diagnosis of Brain Tumors"	http://dx.doi.org/10.1080/0191312049 0515603	"Wet" scanning electron microscopy (wet SEM)	
1998	100	"A Depolarizing Chloride Current Contributes to Chemoelectrical Transduction in Olfactory Sensory Neurons in Situ"	http://www.jneurosci.org/content/18/17/6623.full	Scanning transmission electron microscope (STEM)	
1994	2000	"Enhanced Optical Imaging of Rat Gliomas and Tumor Margins"	http://journals.lww.com/neurosurgery/Abstract/1994/11000/Enhanced_Optical_Imaging_of_Rat_Gliomas_and_Tumor.19.aspx	Enhanced optical imaging	With a spatial resolution of the optical images below 20 microns 2/pixel (22).
1983	3000	"3D Imaging of X-Ray Microscopy"	http://www.scipress.org/e-library/sof2/pdf/0105.PDF	Projection microscopy	See Fig. 7 in article.

23. Spatial resolution in microns (μm) of nondestructive imaging techniques in animals, 1985–2012:

Year	Finding	
2012	Resolution	0.07
	Citation	Sebastian Berning et al., "Nanoscopy in a Living Mouse Brain," *Science* 335, no. 6068 (February 3, 2012): 551.
	URL	http://dx.doi.org/10.1126/science.1215369
	Technique	Stimulated emission depletion (STED) fluorescence nanoscopy
	Notes	Highest resolution achieved in vivo so far
2012	Resolution	0.25
	Citation	Sebastian Berning et al., "Nanoscopy in a Living Mouse Brain," *Science* 335, no. 6068 (February 3, 2012): 551.
	URL	http://dx.doi.org/10.1126/science.1215369
	Technique	Confocal and multiphoton microscopy
2004	Resolution	50
	Citation	Amiram Grinvald and Rina Hildesheim, "VSDI: A New Era in Functional Imaging of Cortical Dynamics," *Nature Reviews Neuroscience* 5 (November 2004): 874–85.
	URL	http://dx.doi.org/10.1038/nrn1536
	Technique	Imaging based on voltage-sensitive dyes (VSDI)
	Notes	"VSDI has provided high-resolution maps, which correspond to cortical columns in which spiking occurs, and offer a spatial resolution better than 50 μm."
1996	Resolution	50
	Citation	Dov Malonek and Amiram Grinvald, "Interactions between Electrical Activity and Cortical Microcirculation Revealed by Imaging Spectroscopy: Implications for Functional Brain Mapping," *Science* 272, no. 5261 (April 26, 1996): 551–54.

Year	Finding		
	URL		http://dx.doi.org/10.1126/science.272.5261.551
	Technique		Imaging spectroscopy
	Notes		"The study of spatial relationships between individual cortical columns within a given brain area has become feasible with optical imaging based on intrinsic signals, at a spatial resolution of about 50 μm."
1995	Resolution		50
	Citation		D. H. Turnbull et al., "Ultrasound Backscatter Microscope Analysis of Early Mouse Embryonic Brain Development," *Proceedings of the National Academy of Sciences* 92, no. 6 (March 14, 1995): 2239–43.
	URL		http://www.pnas.org/content/92/6/2239.short
	Technique		Ultrasound backscatter microscopy
	Notes		"We demonstrate application of a real-time imaging method called ultrasound backscatter microscopy for visualizing mouse early embryonic neural tubes and hearts. This method was used to study live embryos in utero between 9.5 and 11.5 days of embryogenesis, with a spatial resolution close to 50 μm."
1985	Resolution		500
	Citation		H. S. Orbach, L. B. Cohen, and A. Grinvald, "Optical Mapping of Electrical Activity in Rat Somatosensory and Visual Cortex," *Journal of Neuroscience* 5, no. 7 (July 1, 1985): 1886–95.
	URL		http://www.jneurosci.org/content/5/7/1886.short
	Technique		Optical methods

Chapter 11: Objections

1. Paul G. Allen and Mark Greaves, "Paul Allen: The Singularity Isn't Near," *Technology Review,* October 12, 2011, http://www.technologyreview.com/blog/guest/27206/.
2. ITRS, "International Technology Roadmap for Semiconductors," http://www.itrs.net/Links/2011ITRS/Home2011.htm.

3. Ray Kurzweil, *The Singularity Is Near* (New York: Viking, 2005), chapter 2.

4. Endnote 2 in Allen and Greaves, "The Singularity Isn't Near," reads as follows: "We are beginning to get within range of the computer power we might need to support this kind of massive brain simulation. Petaflop-class computers (such as IBM's BlueGene/P that was used in the Watson system) are now available commercially. Exaflop-class computers are currently on the drawing boards. These systems could probably deploy the raw computational capability needed to simulate the firing patterns for all of a brain's neurons, though currently it happens many times more slowly than would happen in an actual brain."

5. Kurzweil, *The Singularity Is Near*, chapter 9, section titled "The Criticism from Software" (pp. 435–42).

6. Ibid., chapter 9.

7. Although it is not possible to precisely determine the information content in the genome, because of the repeated base pairs it is clearly much less than the total uncompressed data. Here are two approaches to estimating the compressed information content of the genome, both of which demonstrate that a range of 30 to 100 million bytes is conservatively high.

1. In terms of the uncompressed data, there are 3 billion DNA rungs in the human genetic code, each coding 2 bits (since there are four possibilities for each DNA base pair). Thus the human genome is about 800 million bytes uncompressed. The noncoding DNA used to be called "junk DNA," but it is now clear that it plays an important role in gene expression. However, it is very inefficiently coded. For one thing, there are massive redundancies (for example, the sequence called "ALU" is repeated hundreds of thousands of times), which compression algorithms can take advantage of.

With the recent explosion of genetic data banks, there is a great deal of interest in compressing genetic data. Recent work on applying standard data compression algorithms to genetic data indicates that reducing the data by 90 percent (for bit perfect compression) is feasible: Hisahiko Sato et al., "DNA Data Compression in the Post Genome Era," *Genome Informatics* 12 (2001): 512–14, http://www.jsbi.org/journal/GIW01/GIW01P130.pdf.

Thus we can compress the genome to about 80 million bytes without loss of information (meaning we can perfectly reconstruct the full 800-million-byte uncompressed genome).

Now consider that more than 98 percent of the genome does not code for proteins. Even after standard data compression (which eliminates redundancies and uses a dictionary lookup for common sequences), the algorithmic content of the noncoding regions appears to be rather low, meaning that it is likely that we could code an algorithm that would perform the same function with fewer bits.

However, since we are still early in the process of reverse-engineering the genome, we cannot make a reliable estimate of this further decrease based on a functionally equivalent algorithm. I am using, therefore, a range of 30 to 100 million bytes of compressed information in the genome. The top part of this range assumes only data compression and no algorithmic simplification.

Only a portion (although the majority) of this information characterizes the design of the brain.

2. Another line of reasoning is as follows. Though the human genome contains around 3 billion bases, only a small percentage, as mentioned above, codes for proteins. By current estimates, there are 26,000 genes that code for proteins. If we assume those genes average 3,000 bases of useful data, those equal only approximately 78 million bases. A base of DNA requires only 2 bits, which translate to about 20 million bytes (78 million bases divided by four). In the protein-coding sequence of a gene, each "word" (codon) of three DNA bases translates into one amino acid. There are, therefore, 4^3 (64) possible codon codes, each consisting of three DNA bases. There are, however, only 20 amino acids used plus a stop codon (null amino acid) out of the 64. The rest of the 43 codes are used as synonyms of the 21 useful ones. Whereas 6 bits are required to code for 64 possible combinations, only about 4.4 ($\log_2 21$) bits are required to code for 21 possibilities, a savings of 1.6 out of 6 bits (about 27 percent), bringing us down to about 15 million bytes. In addition, some standard compression based on repeating sequences is feasible here, although much less compression is possible on this protein-coding portion of the DNA than in the so-called junk DNA, which has massive redundancies. So this will bring the figure probably below 12 million bytes. However, now we have to add information for the noncoding portion of the DNA that controls gene expression. Although this portion of the DNA constitutes the bulk of the genome, it appears to have a low level of information content and is replete with massive redundancies. Estimating that it matches the approximately 12 million bytes of protein-coding DNA, we again come to approximately 24 million bytes. From this perspective, an estimate of 30 to 100 million bytes is conservatively high.

8. Dharmendra S. Modha et al., "Cognitive Computing," *Communications of the ACM* 54, no. 8 (2011): 62–71, http://cacm.acm.org/magazines/2011/8/114944-cognitive-computing/fulltext.

9. Kurzweil, *The Singularity Is Near*, chapter 9, section titled "The Criticism from Ontology: Can a Computer Be Conscious?" (pp. 458–69).

10. Michael Denton, "Organism and Machine: The Flawed Analogy," in *Are We Spiritual Machines? Ray Kurzweil vs. the Critics of Strong AI* (Seattle: Discovery Institute, 2002).

11. Hans Moravec, *Mind Children* (Cambridge, MA: Harvard University Press, 1988).

NOTES

Epilogue

1. "In U.S., Optimism about Future for Youth Reaches All-Time Low," Gallup Politics, May 2, 2011, http://www.gallup.com/poll/147350/optimism-future-youth
-reaches-time-low.aspx.
2. James C. Riley, *Rising Life Expectancy: A Global History* (Cambridge: Cambridge University Press, 2001).
3. J. Bradford DeLong, "Estimating World GDP, One Million B.C.—Present," May 24, 1998, http://econ161.berkeley.edu/TCEH/1998_Draft/World_GDP/Estimating
_World_GDP.html, and http://futurist.typepad.com/my_weblog/2007/07/economic
-growth.html. See also Peter H. Diamandis and Steven Kotler, *Abundance: The Future Is Better Than You Think* (New York: Free Press, 2012).
4. Martine Rothblatt, *Transgender to Transhuman* (privately printed, 2011). She explains how a similarly rapid trajectory of acceptance is most likely to occur for "transhumans," for example, nonbiological but convincingly conscious minds as discussed in chapter 9.
5. The following excerpt from *The Singularity Is Near,* chapter 3 (pp. 133–35), by Ray Kurzweil (New York: Viking, 2005), discusses the limits of computation based on the laws of physics:

> The ultimate limits of computers are profoundly high. Building on work by University of California at Berkeley Professor Hans Bremermann and nanotechnology theorist Robert Freitas, MIT Professor Seth Lloyd has estimated the maximum computational capacity, according to the known laws of physics, of a computer weighing one kilogram and occupying one liter of volume—about the size and weight of a small laptop computer—what he calls the "ultimate laptop."
>
> [Note: Seth Lloyd, "Ultimate Physical Limits to Computation," *Nature* 406 (2000): 1047–54.
>
> [Early work on the limits of computation were done by Hans J. Bremermann in 1962: Hans J. Bremermann, "Optimization Through Evolution and Recombination," in M. C. Yovits, C. T. Jacobi, C. D. Goldstein, eds., *Self-Organizing Systems* (Washington, D.C.: Spartan Books, 1962), pp. 93–106.
>
> [In 1984 Robert A. Freitas Jr. built on Bremermann's work in Robert A. Freitas Jr., "Xenopsychology," *Analog* 104 (April 1984): 41–53, http://www
.rfreitas.com/Astro/Xenopsychology.htm#SentienceQuotient.]
>
> The potential amount of computation rises with the available energy. We can understand the link between energy and computational capacity as follows. The energy in a quantity of matter is the energy associated with each atom (and subatomic particle). So the more atoms, the more energy. As discussed above, each atom can potentially be used for computation. So the

316

more atoms, the more computation. The energy of each atom or particle grows with the frequency of its movement: the more movement, the more energy. The same relationship exists for potential computation: the higher the frequency of movement, the more computation each component (which can be an atom) can perform. (We see this in contemporary chips: the higher the frequency of the chip, the greater its computational speed.)

So there is a direct proportional relationship between the energy of an object and its potential to perform computation. The potential energy in a kilogram of matter is very large, as we know from Einstein's equation $E = mc^2$. The speed of light squared is a very large number: approximately 10^{17} meter^2/second^2. The potential of matter to compute is also governed by a very small number, Planck's constant: 6.6×10^{-34} joule-seconds (a joule is a measure of energy). This is the smallest scale at which we can apply energy for computation. We obtain the theoretical limit of an object to perform computation by dividing the total energy (the average energy of each atom or particle times the number of such particles) by Planck's constant.

Lloyd shows how the potential computing capacity of a kilogram of matter equals pi times energy divided by Planck's constant. Since the energy is such a large number and Planck's constant is so small, this equation generates an extremely large number: about 5×10^{50} operations per second.

[Note: $\pi \times$ maximum energy (10^{17} kg $\times$ meter^2/second^2) / (6.6×10^{-34}) joule-seconds) = $\sim 5 \times 10^{50}$ operations/second.]

If we relate that figure to the most conservative estimate of human brain capacity (10^{19} cps and 10^{10} humans), it represents the equivalent of about 5 billion trillion human civilizations.

[Note: 5×10^{50} cps is equivalent to 5×10^{21} (5 billion trillion) human civilizations (each requiring 10^{29} cps).]

If we use the figure of 10^{16} cps that I believe will be sufficient for functional emulation of human intelligence, the ultimate laptop would function at the equivalent brain power of 5 trillion trillion human civilizations.

[Note: Ten billion (10^{10}) humans at 10^{16} cps each is 10^{26} cps for human civilization. So 5×10^{50} cps is equivalent to 5×10^{24} (5 trillion trillion) human civilizations.]

Such a laptop could perform the equivalent of all human thought over the last ten thousand years (that is, ten billion human brains operating for ten thousand years) in one ten-thousandth of a nanosecond.

[Note: This estimate makes the conservative assumption that we've had ten billion humans for the past ten thousand years, which is obviously not the case. The actual number of humans has been increasing gradually over the past to reach about 6.1 billion in 2000. There are 3×10^7 seconds in a year, and 3×10^{11} seconds in ten thousand years. So, using the estimate of 10^{26} cps for human civilization, human thought over ten thousand years is equivalent to certainly no more than 3×10^{37} calculations. The ultimate laptop performs 5×10^{50} calculations in one second. So simulating ten thousand years of ten billion humans' thoughts would take it about 10^{-13} seconds, which is one ten-thousandth of a nanosecond.]

Again, a few caveats are in order. Converting all of the mass of our 2.2-pound laptop into energy is essentially what happens in a thermonuclear explosion. Of course, we don't want the laptop to explode but to stay within its one-liter dimension. So this will require some careful packaging, to say the least. By analyzing the maximum entropy (degrees of freedom represented by the state of all the particles) in such a device, Lloyd shows that such a computer would have a theoretical memory capacity of 10^{31} bits. It's difficult to imagine technologies that would go all the way in achieving these limits. But we can readily envision technologies that come reasonably close to doing so. As the University of Oklahoma project shows, we already demonstrated the ability to store at least fifty bits of information per atom (although only on a small number of atoms, so far). Storing 10^{27} bits of memory in the 10^{25} atoms in a kilogram of matter should therefore be eventually achievable.

But because many properties of each atom could be exploited to store information—such as the precise position, spin, and quantum state of all of its particles—we can probably do somewhat better than 10^{27} bits. Neuroscientist Anders Sandberg estimates the potential storage capacity of a hydrogen atom at about four million bits. These densities have not yet been demonstrated, however, so we'll use the more conservative estimate.

[Note: Anders Sandberg, "The Physics of the Information Processing Superobjects: Daily Life Among the Jupiter Brains," *Journal of Evolution and Technology* 5 (December 22, 1999), http://www.transhumanist.com/volume5/Brains2.pdf.]

As discussed above, 10^{42} calculations per second could be achieved without producing significant heat. By fully deploying reversible computing techniques, using designs that generate low levels of errors, and allowing for reasonable amounts of energy dissipation, we should end up somewhere between 10^{42} and 10^{50} calculations per second.

The design terrain between these two limits is complex. Examining the technical issues that arise as we advance from 10^{42} to 10^{50} is beyond the scope of this chapter. We should keep in mind, however, that the way this will play out is not by starting with the ultimate limit of 10^{50} and working backward based on various practical considerations. Rather, technology will continue to ramp up, always using its latest prowess to progress to the next level. So once we get to a civilization with 10^{42} cps (for every 2.2 pounds), the scientists and engineers of that day will use their essentially vast nonbiological intelligence to figure out how to get 10^{43}, then 10^{44}, and so on. My expectation is that we will get very close to the ultimate limits.

Even at 10^{42} cps, a 2.2-pound "ultimate portable computer" would be able to perform the equivalent of all human thought over the last ten thousand years (assumed at ten billion human brains for ten thousand years) in ten microseconds.

[Note: See note above. 10^{42} cps is a factor of 10^{-8} less than 10^{50} cps, so one ten-thousandth of a nanosecond becomes 10 microseconds.]

If we examine the Exponential Growth of Computing chart (chapter 2), we see that this amount of computing is estimated to be available for one thousand dollars by 2080.

INDEX

Page numbers in *italics* refer to graphs and illustrations.